Santo Fortunato, Giuseppe Mangioni, Ronaldo Menezes, and Vincenzo Nicosia (Eds.)

Complex Networks

AF788040

Studies in Computational Intelligence, Volume 207

Editor-in-Chief
Prof. Janusz Kacprzyk
Systems Research Institute
Polish Academy of Sciences
ul. Newelska 6
01-447 Warsaw
Poland
E-mail: kacprzyk@ibspan.waw.pl

Further volumes of this series can be found on our homepage: springer.com

Vol. 185. Anthony Brabazon and Michael O'Neill (Eds.)
Natural Computing in Computational Finance, 2009
ISBN 978-3-540-95973-1

Vol. 186. Chi-Keong Goh and Kay Chen Tan
Evolutionary Multi-objective Optimization in Uncertain Environments, 2009
ISBN 978-3-540-95975-5

Vol. 187. Mitsuo Gen, David Green, Osamu Katai, Bob McKay, Akira Namatame, Ruhul A. Sarker and Byoung-Tak Zhang (Eds.)
Intelligent and Evolutionary Systems, 2009
ISBN 978-3-540-95977-9

Vol. 188. Agustín Gutiérrez and Santiago Marco (Eds.)
Biologically Inspired Signal Processing for Chemical Sensing, 2009
ISBN 978-3-642-00175-8

Vol. 189. Sally McClean, Peter Millard, Elia El-Darzi and Chris Nugent (Eds.)
Intelligent Patient Management, 2009
ISBN 978-3-642-00178-9

Vol. 190. K.R. Venugopal, K.G. Srinivasa and L.M. Patnaik
Soft Computing for Data Mining Applications, 2009
ISBN 978-3-642-00192-5

Vol. 191. Zong Woo Geem (Ed.)
Music-Inspired Harmony Search Algorithm, 2009
ISBN 978-3-642-00184-0

Vol. 192. Agus Budiyono, Bambang Riyanto and Endra Joelianto (Eds.)
Intelligent Unmanned Systems: Theory and Applications, 2009
ISBN 978-3-642-00263-2

Vol. 193. Raymond Chiong (Ed.)
Nature-Inspired Algorithms for Optimisation, 2009
ISBN 978-3-642-00266-3

Vol. 194. Ian Dempsey, Michael O'Neill and Anthony Brabazon (Eds.)
Foundations in Grammatical Evolution for Dynamic Environments, 2009
ISBN 978-3-642-00313-4

Vol. 195. Vivek Bannore and Leszek Swierkowski
Iterative-Interpolation Super-Resolution Image Reconstruction: A Computationally Efficient Technique, 2009
ISBN 978-3-642-00384-4

Vol. 196. Valentina Emilia Balas, János Fodor and Annamária R. Várkonyi-Kóczy (Eds.)
Soft Computing Based Modeling in Intelligent Systems, 2009
ISBN 978-3-642-00447-6

Vol. 197. Mauro Birattari
Tuning Metaheuristics, 2009
ISBN 978-3-642-00482-7

Vol. 198. Efrén Mezura-Montes (Ed.)
Constraint-Handling in Evolutionary Optimization, 2009
ISBN 978-3-642-00618-0

Vol. 199. Kazumi Nakamatsu, Gloria Phillips-Wren, Lakhmi C. Jain, and Robert J. Howlett (Eds.)
New Advances in Intelligent Decision Technologies, 2009
ISBN 978-3-642-00908-2

Vol. 200. Dimitri Plemenos, and Georgios Miaoulis
Visual Complexity and Intelligent Computer Graphics Techniques Enhancements, 2009
ISBN 978-3-642-01258-7

Vol. 201. Aboul-Ella Hassanien, Ajith Abraham, Athanasios V. Vasilakos, and Witold Pedrycz (Eds.)
Foundations of Computational Intelligence Volume 1, 2009
ISBN 978-3-642-01081-1

Vol. 202. Aboul-Ella Hassanien, Ajith Abraham, and Francisco Herrera (Eds.)
Foundations of Computational Intelligence Volume 2, 2009
ISBN 978-3-642-01532-8

Vol. 203. Ajith Abraham, Aboul-Ella Hassanien, Patrick Siarry, and Andries Engelbrecht (Eds.)
Foundations of Computational Intelligence Volume 3, 2009
ISBN 978-3-642-01084-2

Vol. 204. Ajith Abraham, Aboul-Ella Hassanien, and André Ponce de Leon F. de Carvalho (Eds.)
Foundations of Computational Intelligence Volume 4, 2009
ISBN 978-3-642-01087-3

Vol. 205. Ajith Abraham, Aboul-Ella Hassanien, and Václav Snášel (Eds.)
Foundations of Computational Intelligence Volume 5, 2009
ISBN 978-3-642-01535-9

Vol. 206. Ajith Abraham, Aboul-Ella Hassanien, André Ponce de Leon F. de Carvalho, and Václav Snášel (Eds.)
Foundations of Computational Intelligence Volume 6, 2009
ISBN 978-3-642-01090-3

Vol. 207. Santo Fortunato, Giuseppe Mangioni, Ronaldo Menezes, and Vincenzo Nicosia (Eds.)
Complex Networks, 2009
ISBN 978-3-642-01205-1

Santo Fortunato, Giuseppe Mangioni,
Ronaldo Menezes, and Vincenzo Nicosia (Eds.)

Complex Networks

Results of the 2009 International Workshop on Complex Networks (CompleNet 2009)

Springer

Santo Fortunato
Complex Networks Lagrange Laboratory
ISI Foundation
Viale S. Severo 65
10133 Torino
Italy
E-mail: fortunato@isi.it

Giuseppe Mangioni
Dipartimento di Ingegneria
Informatica e delle Telecomunicazioni
Università degli Studi di Catania
Viale Andrea Doria, 6
I95125 Catania
Italy
E-mail: Giuseppe.Mangioni@diit.unict.it

Ronaldo Menezes
Florida Institute of Technology
Department of Computer Sciences
150 W. University Blvd
Melbourne FL 32901
USA
E-mail: rmenezes@cs.fit.edu

Vincenzo Nicosia
Laboratorio sui Sistemi Complessi
Scuola Superiore di Catania
Via S. Nullo 5/i – 95123 Catania
Italy
E-mail: vincenzo.nicosia@ct.infn.it

ISBN 978-3-642-26931-8 e-ISBN 978-3-642-01206-8

DOI 10.1007/978-3-642-01206-8

Studies in Computational Intelligence ISSN 1860949X

© 2009 Springer-Verlag Berlin Heidelberg
Softcover reprint of the hardcover 1st edition 2009

This work is subject to copyright. All rights are reserved, whether the whole or part of the material is concerned, specifically the rights of translation, reprinting, reuse of illustrations, recitation, broadcasting, reproduction on microfilm or in any other way, and storage in data banks. Duplication of this publication or parts thereof is permitted only under the provisions of the German Copyright Law of September 9, 1965, in its current version, and permission for use must always be obtained from Springer. Violations are liable to prosecution under the German Copyright Law.

The use of general descriptive names, registered names, trademarks, etc. in this publication does not imply, even in the absence of a specific statement, that such names are exempt from the relevant protective laws and regulations and therefore free for general use.

Typeset & Cover Design: Scientific Publishing Services Pvt. Ltd., Chennai, India.

Printed in acid-free paper

9 8 7 6 5 4 3 2 1

springer.com

Editorial

Santo Fortunato and Ronaldo Menezes

Complex Networks

Complexity science is a major scientific revolution of our time. A system is complex if its macroscopic behavior cannot be simply reduced to the microscopic behavior of its constituents. Nonlinear dynamics and feedback effects are the dominant characters of the dynamics of complex systems. A major contribution of the last years is the discovery that many complex systems in biology, physics, social and computer sciences, can be studied through their graph (network) representation, in which the elementary units of a system become nodes and their interactions links connecting nodes [1, 2, 3, 5, 4, 6]. Examples are social networks of acquaintances, protein-protein interaction networks, food webs, the Internet and the World Wide Web. Networks representing real systems are characterized by a set of common features, which markedly differ from features of simple lattices or random graphs. This is why one refers to them as *complex networks*. The most striking feature is the fact that the distribution of the number of neighbors of a node (degree) is typically broad, with a tail that often follows a power law [7]. This means that nodes with low degree coexist with nodes with large degree, or hubs. The presence of hubs is responsible for a number of remarkable features of complex networks, like their resilience against random attacks or failures [8], and the absence of percolation and epidemic thresholds [8, 9]. Another important characteristic of complex networks concerns clustering, i.e. the probability that neighbors of a node are themselves neighbors [10]. Real networks usually display high values for this probability, much higher than in random graphs, and this property plays an important role in spreading

Santo Fortunato
ISI Foundation, Viale S. Severo 65, 10133 Torino, Italy.
e-mail: fortunato@isi.it

Ronaldo Menezes
Bio-Inspired Computing Lab, Department of Computer Sciences, Florida Institute of Technology, Melbourne, Florida, USA.
e-mail: rmenezes@cs.fit.edu

phenomena. Finally, complex networks are characterized by short geodesic paths between nodes, i.e. it is possible to reach every node from any other node in a small number of steps, that grows only logarithmically with the size of the system (small-world property) [10]. The small-world property holds as well for simple random graphs, but in complex networks geodesic paths are much shorter, due to the presence of the hubs, which are fundamental communication stations.

The main issues in the theory of complex networks concern their *structure* and their *dynamics*. A lot of work has been devoted to describe the structure of complex networks. Local properties like degree, clustering coefficient, node centrality, etc. have been studied, along with their statistical distributions. Moreover, complex networks display interesting structures at the mesoscopic level, consisting in groups of nodes with many links between nodes of the same group and comparatively few between nodes of different groups [11]. Such groups of nodes are called *communities*, and represent an essential aspect of the organization of networks, which allow to infer nontrivial relationships between the nodes.

As far as the dynamics is concerned, one has to distinguish the dynamics of complex networks, from the dynamics of processes taking place on them. Models of network dynamics have been proposed since the origins of complex networks. They are mostly models of growth, in which nodes are iteratively added to a small initial core. The fat-tailed distributions of degree observed in real systems are produced by the tendency of nodes to set links with nodes with large degree [12], although several other mechanisms have been proposed. Studying processes taking place on networks is essential to understand how networks function. Many traditional processes, like percolation, spin models dynamics, synchronization, opinion dynamics, epidemic spreading, etc. have been thoroughly analyzed, highlighting the role of hubs as fundamental actors in such processes.

Nowadays network theory is rather developed. Still, there are several open problems and the research activity in this field remains intense. For instance, the problem of detecting communities in networks is still far from being solved and it is now probably the most active research topic in the area. Similarly, a lot of effort has been put in the study of *weighted networks*, where one is confronted with the issue of the interplay between the strength of the interactions and the network topology. Finally, a recent research direction focuses on the issue of the *coevolution* of networks and dynamics, i.e. with the issue of the interplay between the evolution of the network and that of dynamical processes taking place on it, when both evolutions occur on comparative time scales.

This volume aims at increasing our knowledge on network theory and applications by discussing some very recent results in the field. The enclosed papers cover many of the issues above mentioned, and were contributed by a widely interdisciplinary community of scientists, including physicists, mathematicians, computer sciences, engineers and economists. We distinguish five subject areas for the 18 papers of this volume: analysis of real networks (3 papers), community structure (4 papers), network modeling (3 papers), network dynamics (3 papers) and applications (5 papers).

Analysis of Real Networks

The paper by Fagiolo, Reyes and Schiavo deals with the International Trade Network, i.e. the network of import-export relationships between countries. The authors analyzed the most important statistics of network properties in a time span of twenty years and found that the node-statistic distributions are stable, whereas the distribution of positive link weights is turning from a lognormal to a power law. The paper by Nunnari, Puglisi, Bonforte and Spata presents a study of a network built from the analysis of planetary correlations between the activities of volcanos over the last two thousand years. The network displays both the small-world property and high clustering coefficient. Last, the paper by Živković, Mitrović and Tadić focuses on the network derived from correlations in the time series of gene expressions of yeast. Modules of genes can be identified by analyzing the eigenvectors of the adjacency matrix, which is obtained from the correlation matrix through standard filtering.

Community Structure

The paper by Cristino, Andrade and Costa, like paper by Fagiolo *et al.*, deals with the gene interactions of yeast. The goal is to identify the modular structure in gene transcription networks. Communities are found with the modularity optimization algorithm by Clauset *et al.* [13]. Structurally similar communities tend to share functional properties. The paper by Gregory proposes a new method to find overlapping communities in networks. It consists of two steps: first, the network is transformed in a new one where nodes shared between modules are split, i.e. represented in multiple copies; second, a traditional community detection method is applied to the resulting network. For the second phase Gregory adopted several algorithms, based on modularity optimization and random walks. The two-step strategy yields more accurate results than specialized methods to find overlapping communities, like the C-finder by Palla *et al.* [14]. In next paper, Fiumicello, Longheu and Mangioni introduce a fast implementation of a community detection method recently proposed by Lancichinetti *et al.* [15], which consists in the local optimization of a fitness function with the possibility to find overlaps between clusters. The gain in speed is obtained by porting the original algorithm on a grid environment. In the last paper in this group, Xiang, Chen and Zhou introduced an agglomerative community detection method, based on a measure quantifying the structural similarity between graphs. The procedure is much faster than the fast method of Clauset *et al.* [13], based on greedy modularity optimization. A hybrid strategy, combining the new method with the method by Clauset *et al.* yields better estimates for the maximum modularity of networks.

Network Modeling

The first paper in this group, by Brandes, Lerner, Nagel and Nick, introduces a method to identify structural trends in network ensembles. The ensembles are mixtures of planted partition models, i.e. classes of graphs with well defined groups of nodes. The method relies on the relative comparison of the vectors whose components are the eigenvalues of the adjacency matrix of graph instances of the ensembles. The paper by Gustedt proposes models that generate graphs with high clustering coefficient. The models are a generalization of the recipes behind the construction of the random graphs *á la* Erdös and Rényi [16] and scale-free networks *á la* Barabási-Albert [12]. The paper by Villas Boas, Rodrigues and Costa suggests a model of highway networks, which generalizes the concept of geographical networks, in that new nodes are introduced between pairs of important nodes, which simulates the presence of small intermediate towns between important centers. The model is able to describe the US highway network better than traditional complex networks models.

Network Dynamics

This group starts with the paper by Maletić and Rajković where they introduce a general model for opinion dynamics that combines two other popular dynamics. At variance with traditional models, here new opinions are formed. The dynamics runs on different types of scale-free networks, and the final opinion state strongly depends on whether the network has a modular structure or not. The paper by Loya and Lucas proves analytically results on the stationary state of a general epidemic model, including as special cases popular models like the SIS (susceptible-infected-susceptible) model. The dynamics runs on networks with degree-degree correlations. The epidemic threshold is also rigorously defined. The paper authored by Melnik and Gleeson, presents an analytical approach to determine the thresholds and the sizes of the giant components for bond percolation on random networks with non-zero clustering. This study suggests that, for a network with a given level of clustering, the percolation threshold may be higher or lower than the threshold on a randomized (unclustered) network with the same degree distribution, depending on the degree distribution.

Applications

The first paper of the application group, authored by Funabashi, Chavalarias and Cointet, studied the correlations between variables on network nodes beyond the classic dyadic interactions. They use information geometry to decompose correlations into different orders of statistics. High order correlations are used to analyze networks of word occurrences derived from political weblog data. In the paper by Hoche, Hardcastle and Flach, they introduce a method to reduce the computational complexity of topic prediction in co-authorship graphs. The method consists in reducing the number of links of the graph, by using the publication dates of the papers. Tests on real datasets show that the method is able to recover the same predictions as for the full graph, even when a large fraction of links is removed, with a gain in computational efficiency of one order of magnitude. The group follow with the paper by Aoyama, Saito, Yamada and Ueda, which presents a graph-based approach for a fast search of the most similar object to a given query object. The approach is based on the fact that small-world networks are "searchable", in that any node can reach any target node using only local information. The similarity search is turned into a search problem on the similarity network of objects. The method performs well with the gain of a low expected search cost. The paper by Lawniczak, Wu and Di Stefano deals with the problem of network security, by focusing on the detection of anomalous traffic packets in networks like the Internet. They use information entropy to monitor packet traffic, and find that the entropy profiles of packet traffic at selected nodes/routers may indeed give information on the traffic behavior over the whole network, and reveal anomalies due to distributed denial-of-service attacks. Finally, in the paper by Collingsworth and Menezes, a temporal analysis of the email network of Enron employees is performed, with the goal of identifying properties of the network that anticipate the occurrence of social tensions between people. In particular, they find that an increase in the number of tightly connected groups seems to be a signal of social tension.

Acknowledgements. We would like to thank Springer for the publication of this volume. We are also very grateful to all the authors of papers contained in the volume and to the members of CompleNet 2009 Program Committee, whose activity has been invaluable to guarantee a careful screening of the submitted contributions.

References

1. Albert, R., Barabási, A.-L.: Statistical mechanics of complex networks. Rev. Mod. Phys. 74, 47–97 (2002)
2. Mendes, J.F.F., Dorogovtsev, S.N.: Evolution of Networks: from biological nets to the Internet and WWW. Oxford University Press, Oxford (2003)
3. Newman, M.E.J.: The structure and function of complex networks. SIAM Rev. 45, 167–256 (2003)
4. Pastor-Satorras, R., Vespignani, A.: Evolution and structure of the Internet: a statistical physics approach. Cambridge University Press, Cambridge (2004)
5. Boccaletti, S., Latora, V., Moreno, Y., Chavez, M., Hwang, D.-U.: Complex networks: Structure and dynamics. Phys. Rep. 424, 175–308 (2006)
6. Barrat, A., Barthélemy, M., Vespignani, A.: Dynamical Processes on Complex Networks. Cambridge University Press, Cambridge (2008)
7. Albert, R., Jeong, H., Barabási, A.-L.: The diameter of the World Wide Web. Nature 401, 130–131 (1999)
8. Cohen, R., Erez, K., ben Avraham, D., Havlin, S.: Resilience of the Internet to Random Breakdown. Phys. Rev. Lett. 85, 4626–4628 (2000)
9. Pastor-Satorras, R., Vespignani, A.: Epidemic spreading in scale-free networks. Phys. Rev. Lett. 86, 3200–3203 (2001)
10. Watts, D.J., Strogatz, S.H.: Collective dynamics of "small-world" networks. Nature 393, 440–442 (1998)
11. Girvan, M., Newman, M.E.J.: Community structure in social and biological networks. Proc. Natl. Acad. Sci. USA 99, 7821–7826 (2002)
12. Barabási, A.-L., Albert, R.: Emergence of scaling in random networks. Science 286, 509–512 (1999)
13. Clauset, A., Newman, M.E.J., Moore, C.: Phys. Rev. E 70, 066111 (2004)
14. Palla, G., Derényi, I., Farkas, I., Vicsek, T.: Nature 435, 814–818 (June 9, 2005)
15. Lancichinetti, A., Fortunato, S., Kertész, J.: Detecting the overlapping and hierarchical community structure of complex networks Eprint arXiv: 0802.1218
16. Erdös, P., Rényi, A.: On random graphs I. Publ. Math. Debrecen 6, 290–297 (1959)

CompleNet
Workshop on Complex Networks

Welcome to Catania, Italy, for the first International Workshop on Complex Networks (CompleNet 2009). This workshop series was an initiative of research groups from: Dipartimento di Ingegneria Informatica e delle Telecomunicazioni (DIIT) - University of Catania - Italy, the Laboratorio sui Sistemi Complessi – Scuola Superiore di Catania, Italy, and the Bio-Inspired Computing Laboratory in the Department of Computer Sciences, Florida Institute of Technology, USA. CompleNet 2009 was hosted by the DIIT, Engineering Faculty, University of Catania, Italy during May 26–27, 2009.

CompleNet aims at bringing together researchers and practitioners working on areas related to complex networks into an intensive 2 days high quality program. Despite the existence of other meetings, this is the first attempt to bring Complex Network researchers into a format long known by the Computer Science and Engineering communities but relatively uncommon for the Sciences (Physics, Biology, Social Sciences, etc.). The main difference lies on the fact that CompleNet requires full papers to be submitted and not only abstracts. We believe this improves the quality of the workshop and this book is proof of this.

In the past two decades we have been witnessing an exponential increase on the number of publications in Complex Networks. From biological systems to computer science, from economic to social systems, complex networks are becoming pervasive in many fields of science. It is this interdisciplinary nature of complex networks that this workshop aims at addressing as we believe there is still a long fruitful way for us to explore. The exchange of ideas between the areas helps in the acceleration of this process.

We would like to thank first all the authors and participants of the workshop; they are the main reason this event was a sucess. Putting a workshop together requires a lot of work and dedication particularly when we wanted to make sure the results (this book) would stand on its own. We are particularly grateful to the Program Chairs Santo Fortunato and Ronaldo Menezes for their many contributions and extensive work towards putting together a very successful program. We would like to thank the members the Program Committee for their dedication and invaluable expert reviews, ensuring a high-quality technical program. We are also particularly grateful to the keynote speakers, José Fernando Mendes and Alessandro Vespignani, for taking the time to participate in this workshop and to share their expertise with us. Last but certainly not least, we would like to give our special thanks to Prof. Vincenza Carchiolo, Prof. Michele Malgeri and Dr. Alessandro Longheu from the DIIT and to those people who have kindly helped us prepare and organize the CompleNet 2009 workshop.

We hope to see you in future editions of this workshop and invite you to consider submitting your work in 2010 and beyond.

Catania, Italy, May 2009
CompleNet Steering Committee

Giuseppe Mangioni
Ronaldo Menezes
Vincenzo Nicosia

Conference Organization

Steering Committee

Giuseppe Mangioni
Ronaldo Menezes
Vincenzo Nicosia

Program Chairs

Santo Fortunato
Ronaldo Menezes

Program Committee

Alex Arenas
Alain Barrat
Marc Barthelemy
Ginestra Bianconi
Haluk Bingol
Marian Boguna
Stefan Bornholdt
Dan Braha
Angelo Brayner
Guido Caldarelli
Vincenza Carchiolo
Claudio Castellano
Aaron Clauset
Luciano Costa
Paolo De Los Rios
Gustavo Deco
Kwang-Il Goh

Roger Guimera
Kimmo Kaski
Alessandro Longheu
Vittorio Loreto
Adilson Motter
Akira Namatame
Sidney Redner
Eraldo Ribeiro
M. Ãngeles Serrano
Igor Sokolov
Bosiljka Tadic
Andres Upegui
Soon-Hyung Yook

Additional Reviewers

Fragkiskos Papadopoulos

Local Organization

Vincenza Carchiolo
Alessandro Longheu
Michele Malgeri
Giuseppe Mangioni
Vincenzo Nicosia

Contents

Dynamics and Evolution of the International Trade Network*

Giorgio Fagiolo, Javier Reyes, and Stefano Schiavo

Abstract. This paper studies how the distributions of the most important network statistics measuring connectivity, assortativity, clustering and centrality in the international-trade network have co-evolved over time. We show that all node-statistic distributions and their correlation structure have remained surprisingly stable in the last 20 years – and are likely to do so in the future. Conversely, the distribution of (positive) link weights is slowly moving from a log-normal density towards a power law.

1 Introduction

The last years have witnessed the emergence of a large body of contributions addressing international-trade issues from a complex-network perspective (Li et al., 2003; Serrano and Boguñá, 2003; Garlaschelli and Loffredo, 2004,

Giorgio Fagiolo
Sant'Anna School of Advanced Studies, Laboratory of Economics and Management, Piazza Martiri della Libertà 33, I-56127 Pisa, Italy
Tel: +39-050-883356 Fax: +39-050-883343
e-mail: giorgio.fagiolo@sssup.it

Javier Reyes
Department of Economics, Sam M. Walton College of Business, University of Arkansas, USA
e-mail: JReyes@walton.uark.edu

Stefano Schiavo
Department of Economics and School of International Studies, University of Trento, Italy and OFCE, France
e-mail: stefano.schiavo@ofce.sciences-po.fr

* Thanks to two anonymous referees for their helpful comments. This work is an abridged version of the paper: Fagiolo, G., Reyes, J. and Schiavo, S. (2009), "The World-Trade Web: Topological Properties, Dynamics, and Evolution", forthcoming in *Physical Review E*.

R. Menezes et al. (Eds.): Complex Networks: CompleNet 2009, SCI 207, pp. 1–13.
springerlink.com © Springer-Verlag Berlin Heidelberg 2009

2005; Ausloos and Lambiotte, 2006; Miskiewicz and Ausloos, 2006; Gorski et al, 2006; Serrano et al., 2007; Bhattacharya et al., 2007a,b; Garlaschelli et al., 2007; Fagiolo, 2007; Fagiolo et al., 2007a; Tzekina et al., 2007; Fagiolo et al., 2008; Gligor and Ausloos, 2008). The International Trade Network (ITN), is defined as the network of import/export relationships between world countries in a given year. Understanding the topological properties of the ITN, and their evolution over time, acquires a fundamental importance in explaining international-trade issues such as economic globalization and internationalization, as trade linkages are one of the most important channels of interaction between world countries.

The first stream of studies that have exlpored the properties of the ITN has employed a *binary-network analysis*, where a (possibly directed) link between any two countries is either present or not according to whether the trade flow that it carries is larger than a given lower threshold (Serrano and Boguñá, 2003; Garlaschelli and Loffredo, 2004, 2005). More recently, however, a few contributions (Li et al., 2003; Bhattacharya et al., 2007a,b; Garlaschelli et al., 2007; Fagiolo et al., 2008) have adopted a *weighted-network approach* (Barrat et al., 2004; Barthélemy et al., 2005) to the study of the ITN, where each link is weighted by some proxy of the trade intensity that it carries. The motivation is that a binary approach cannot fully extract the wealth of information about the trade intensity flowing through each link and therefore might dramatically underestimate the role of heterogeneity in trade linkages. Indeed, Fagiolo (2007); Fagiolo et al. (2007a, 2008) show that the statistical properties of the ITN viewed as a weighted network crucially differ from those exhibited by its weighted counterpart, and that a weighted-network analysis is able to provide a more complete and truthful picture of the ITN than a binary one.

In this paper, we employ a weighted network approach to characterize, for the period 1981-2000, the main properties governing the dynamics and evolution of the distributions of the most important network statistics measuring node connectivity, assortativity, clustering and centrality; as well as link weights. We ask two main types of questions: (i) Have (and, if so, how) the distributional properties of these statistics (and their correlation structure) been changing within the sample period considered? (ii) Can we make any prediction on the out-of-sample evolution of such distributions?

The rest of the paper is organized as follows. Section 2 presents the data sets and defines the statistics studied in the paper. Section 3 introduces the main results. Finally, Section 4 concludes.

2 Data and Definitions

We employ international-trade data provided by Gleditsch (2002) to build a time-sequence of weighted directed networks. Our balanced panel refers to $T = 20$ years (1981-2000) and $N = 159$ countries. For each country and year, data report trade flows in current US dollars. To build adjacency and weight matrices, we followed the flow of goods. This means that rows represent exporting countries, whereas columns stand for importing countries. We define a "trade relationship" by setting the generic entry of the (binary) adjacency matrix $\tilde{a}^t_{ij} = 1$ if and only if exports from country i to country j (e^t_{ij}) are strictly positive in year t.

Following Li et al. (2003); Bhattacharya et al. (2007a,b); Garlaschelli et al. (2007), the weight of a link from i to j in year t is defined as $\tilde{w}^t_{ij} = e^t_{ij}$. Thus, the sequence of $N \times N$ adjacency and weight matrices $\{\tilde{A}^t, \tilde{W}^t\}$, $t = 1981, ..., 2000$ fully describes the within-sample dynamics of the ITN.

A preliminary statistical analysis of both binary and weighted matrices suggests that $(\tilde{A}^t, \tilde{W}^t)$ are sufficiently symmetric to justify an undirected analysis for all t. From a binary perspective, the majority of ITN links are reciprocated: on average, about 93% of countries export to partners that in turn export to them. To check more formally this evidence from a weighted perspective, we have computed the weighted symmetry index defined in Fagiolo (2006). The index ranges in the sample period between 0.0017 and 0.0043, signalling a strong and stable symmetry of ITN weight matrices[1]. We have therefore symmetrized the network by defining the entries of the new adjacency matrix A^t so that $a^t_{ij} = 1$ if and only if either $\tilde{a}^t_{ij} = 1$ or $\tilde{a}^t_{ji} = 1$, and zero otherwise. Accordingly, the generic entry of the new weight matrix W^t is defined as $w^t_{ij} = \frac{1}{2}(\tilde{w}^t_{ij} + \tilde{w}^t_{ji})$. This means that the symmetrized weight of link ij is proportional to the total trade (imports plus exports) flowing through that link in a given year. Finally, in order to have $w^t_{ij} \in [0,1]$ for all (i,j) and t, we have re-normalized all entries in W^t by their maximum value $w^t_* = max^N_{i,j=1} w^t_{ij}$.

For each $(\tilde{A}^t, \tilde{W}^t)$, we study the distributions of the following node statistics:

- *Node degree* (Albert and Barabási, 2002), defined as $ND^t_i = A^t_{(i)}\mathbf{1}$, where $A^t_{(i)}$ is the i-th row of A^t and $\mathbf{1}$ is a unary vector. ND is a measure of binary connectivity, as it counts the number of trade partners of any given node. Although we mainly focus here on a weighted-network approach, we study ND because of its natural interpretation in terms of number of trade partnerships and bilateral trade agreements.
- *Node strength* (DeMontis et al., 2005), defined as $NS^t_i = W^t_{(i)}\mathbf{1}$, where again $W^t_{(i)}$ is the i-th row of W^t. While ND tells us how many partners a node holds, NS is a measure of weighted connectivity, as it gives us an idea of how intense existing trade relationships of country i are.
- *Node average nearest-neighbor strength* (DeMontis et al., 2005), that is $ANNS^t_i = (A^t_{(i)} W^t \mathbf{1})/(A^t_{(i)}\mathbf{1})$. ANNS measures how intense are trade relationships maintained by the partners of a given node. Therefore, the correlation between ANNS and NS is a measure of network assortativity (if positive) or disassortativity (if negative). It is easy to see that ANNS boils down to average nearest-neighbor degree (ANND) if W^t is replaced by A^t.

[1] The expected value of the statistic in a random graph where link weights are uniformly and independently distributed as a uniform in the unit interval is 0.5 (Fagiolo, 2006). Furthermore, the expected value computed by randomly reshuffling in each year the empirically-observed weights among existing links ranges in the same period from 0.0230 to 0.0410. Therefore, the empirical value is significantly smaller than expected.

- *Weighted clustering coefficient* (Saramaki et al., 2007; Fagiolo, 2007), defined as

$$WCC_i^t = \frac{([W^t]^{[\frac{1}{3}]})_{ii}^3}{ND_i^t(ND_i^t - 1)}.$$

  Here Z_{ii}^3 is the i-th entry on the main diagonal of $Z \cdot Z \cdot Z$ and $Z^{[\frac{1}{3}]}$ stands for the matrix obtained from Z after raising each entry to $1/3$. WCC measures how much clustered a node i is from a weighted perspective, i.e. how much intense are the linkages of trade triangles having country i as a vertex. Again, replacing W^t with A^t, one obtains the standard binary clustering coefficient (BCC), which counts the fraction of triangles existing in the neighborhood of any give node (Watts and Strogatz, 1998).
- *Random-walk betweenness centrality* (Newman, 2005; Fisher and Vega-Redondo, 2006), which is a measure of how much a given country is globally-central in the ITN. A node has a higher random-walk betweenness centrality (RWBC) the more it has a position of strategic significance in the overall structure of the network. In other words, RWBC is the extension of node betweenness centrality (Scott, 2000) to weighted networks and measures the probability that a random signal can find its way through the network and reach the target node where the links to follow are chosen with a probability proportional to their weights.

The above statistics allow one to address the study of node characteristics in terms of four dimensions: connectivity (ND and NS), assortativity (ANND and ANNS, when correlated with ND and NS), clustering (BCC and WCC) and centrality (RWBC). In what follows, we will mainly concentrate the analysis on ND and the other weighted statistics (NS, ANNS, WCC, RWBC), but we occasionally discuss, when necessary, also the behavior of ANND and BCC. We further explore the network-connectivity dimension by studying the time-evolution of the link-weight distribution $w^t = \{w_{ij}^t, i \neq j = 1, \ldots, N\}$.

3 Results

3.1 Within-Sample Moment Dynamics

We begin by studying the within-sample dynamics of the moments of the distributions of node and link statistics and their dynamics. As already found in Bhattacharya et al. (2007a,b), link weight distributions display remarkably stable moments (see Figure 1). The fact that the first four moments of the distribution are fairly constant in the sample period hints to a strong persistence of trade relationships. We shall study this issue in more details below.

A similar stable pattern is detected also for the moments of the distributions of all node statistics under analysis, see Figure 2 for the case of NS distributions. Note that the log-scale on the y-axis may hide the possible presence of trends or cyclical behaviors in the within-sample moment dynamics. Nevertheless, more in-depth

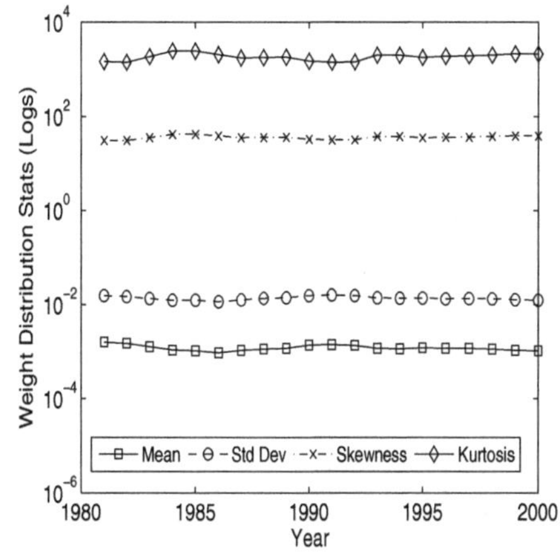

Fig. 1 Sample moments of the link-weight distribution vs. years. Log scale on Y-axis

analyses do not hint to estatistically-significant patterns. This may also be due to the small number of time periods that we study in this paper.

To further explore the topological properties of the ITN, we turn now to examine the dynamics of the correlation structure existing between binary- and weighted-network statistics. Figure 3 shows that countries holding many trade partners and/or very intense trade relationships are also the richest and most (globally) central; typically trade with many countries, but very intensively with only a few very-connected ones; and form few, but intensive, trade clusters (triangles).

Another interesting issue to explore concerns the extent to which country specific characteristics relate to network properties. We focus here on the correlation patterns between network statistics and country per capita GDP (pcGDP) in order to see whether countries with a higher income are more/less connected, central and clustered. The outcomes are very clear and tend to mimic those obtained above for the correlation structure among network statistics. As Figure 4 shows, high-income

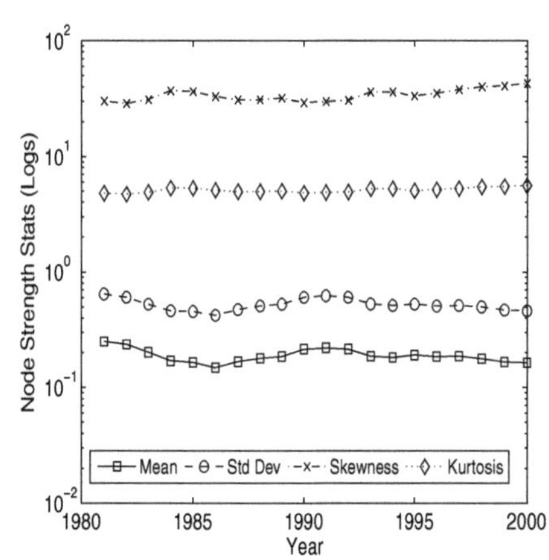

Fig. 2 Sample moments of the node strength (NS) distribution vs. years. Log scale on Y-axis

countries tend to hold more, and more intense, trade relationships and to occupy a more central position. However, they trade with few and weakly-connected partners, a pattern suggesting the presence of a sort of "rich club phenomenon". Again, all empirical correlations are in absolute values larger than their expected values under reshuffling schemes B-RS and W-RS except for the ANNS-pcGDP correlation.

Furthermore, the correlation structure is only weakly changing during the sample period. This suggests that the whole architecture of the ITN has remained fairly stable between 1981 and 2000. To further explore the implications of this result, also in the light of the ongoing processes of internationalization and globalization, we turn now to a more in-depth analysis of the in-sample dynamics and out-of-sample evolution of ITN topological structure.

3.2 *Within-Sample Distribution Dynamics*

The foregoing evidence suggests that the shape of the distributions concerning the most important topological properties of ITN displays a rather strong stability in the 1981-2000 period. However, distributional stability does not automatically rule out the possibility that between any two consecutive time steps, say $t-\tau$ and t, a lot of shape-preserving turbulence was actually going on at the node and link level, with many countries and/or link weights moving back and forth across the quantiles of the distributions. In order to check whether this is the case or not, we have computed stochastic-kernel estimates (Chung, 1960; Futia, 1982) for the distribution dynamics concerning node and link statistics. More formally, consider a real-valued node or link statistic X. Let $\phi^\tau(\cdot,\cdot)$ be the joint distribution of $(X^t, X^{t-\tau})$ and $\psi^\tau(\cdot)$ be the marginal distribution of $X^{t-\tau}$. We estimate the τ-year stochastic kernel, defined as the conditional density $s^\tau(x|y) = \phi^\tau(x,y)/\psi^\tau(y)$.

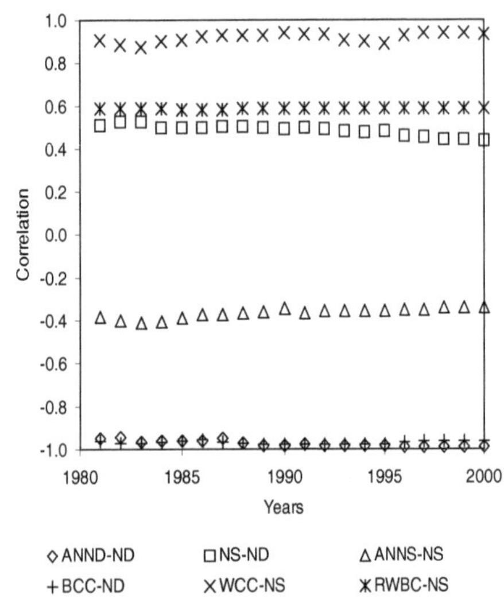

Fig. 3 Correlation between node statistics vs. years

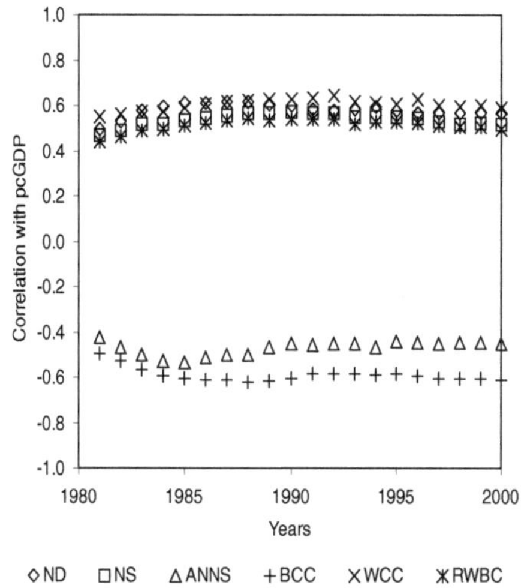

Fig. 4 Correlation between node statistics and country per-capita gross-domestic product (GDP) vs. years

Figures 5 and 6 present the contour plots of the estimates of the 1-year kernel density of logged NS and logged positive link-weights. Notice that the bulk of the probability mass is concentrated close to the main diagonal (displayed as a solid 45° line). Similar results are found for all other real-valued node statistics (ANNS, WCC and RWBC) also at larger time lags. The kernel density of logged positive link weights, contrary to the logged NS one, is instead extremely polarized towards the extremes of the distribution range, whereas in the middle of the range it is somewhat flatter (Figure 6). We will go back to the implications that this feature has on out-of-sample distributional evolution below.

3.3 *Country-Ranking Dynamics*

The distributional-stability results obtained in the foregoing sections naturally hint to the emergence of a lot of stickiness in country rankings (in terms of node

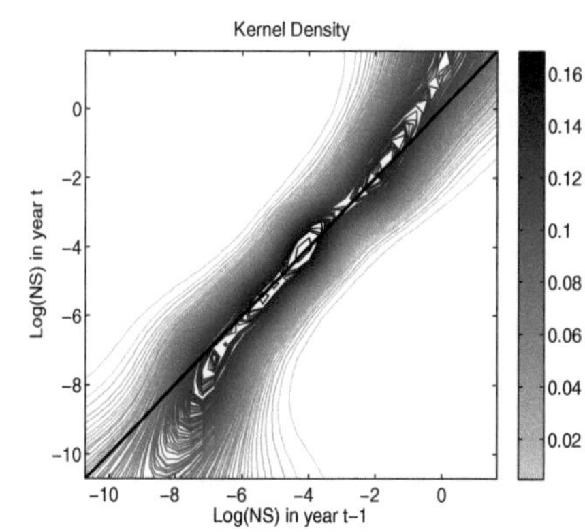

Fig. 5 Contour plot of stochastic-kernel estimates for logged node strength (NS). Solid line: Main 45° diagonal

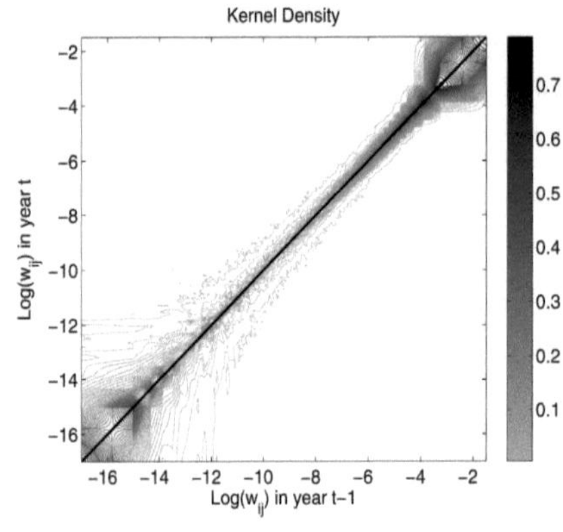

Fig. 6 Contour plot of stochastic-kernel estimates for logged positive link-weights ($\log(w_{ij}^t), w_{ij}^t > 0$). Solid line: Main 45° diagonal

statistics) as well. To explore this issue, for each year $t = 1981, \dots 2000$ we have ranked our $N = 159$ countries according to any of the five main statistics employed so far (ND, NS, ANNS, WCC, RWBC) in a descending order. The first question we are interested in is assessing to which extent also these rankings are sticky across time. We check stability of rankings by computing the time-average of Spearman rank-correlation coefficients (SRCC) (Spearman, 1904; Hollander and Wolfe, 1973) between consecutive years[2]. More formally, let $r_{(i)}^t(X)$ be the rank of country $i = 1, \dots, N$ in year t according to statistic X, and $\rho_{t-1,t}(X)$ be the SRCC between rankings at two consecutive years $t-1$ and t, for $t = 2, \dots, T$. Our ranking-stability index (RSI) for the statistic X is defined as

$$RSI(X) = \frac{1}{T-1} \sum_{t=2}^{T} \rho_{t-1,t}(X). \tag{1}$$

Of course $RSI(X) \in [-1, 1]$, where $RSI(X) = -1$ implies the highest ranking turbulence, whereas $RSI(X) = 1$ indicates complete stability. The results for the ITN suggest that even rankings are very stable over time. Indeed, one has that $RSI(ND) = 0.9833$, $RSI(NS) = 0.9964$, $RSI(ANNS) = 0.9781$, $RSI(WCC) = 0.9851$ and $RSI(RWBC) = 0.9920$. Notice that, since $\rho_{t-1,t}(X) \to N(0, N^{-1})$, our $RSI(X)$ should tend to a $N(0, [N(T-1)]^{-1}) \cong N(0, 3.3102 \times 10^{-4})$. Therefore, our empirical values are more than 50 standard deviations to the right of 0 (no average rank correlation)[3].

The second issue that deserves a closer look concerns detecting which countries rank high according to different statistics. Table 1 displays the top-20 countries in

[2] We focus here only on one-year lags between rankings. An interesting extension to the present analysis would be to check for stability of rankings across time lags of length $\tau > 1$.

[3] An interesting extension of the rank-stability analysis performed here would imply to test for stability of top-ranking countries, or more generally subsets of the entire sample of countries (Braha and Bar-Yam, 2006). This may help one in better assessing the causes and nature of the observed overall ranking stability.

Table 1 Country rankings in year 2000 according to node statistics

Rank	ND	NS	ANNS	WCC	RWBC
1	Germany	USA	Sao Tom & Principe	USA	USA
2	Italy	Germany	Kiribati	Germany	Germany
3	UK	Japan	Nauru	Japan	Japan
4	France	France	Tonga	UK	France
5	Switzerland	China	Vanuatu	China	UK
6	Australia	UK	Tuvalu	France	China
7	Belgium	Canada	Burundi	Italy	Italy
8	Netherlands	Italy	Botswana	Netherlands	S. Korea
9	Denmark	Netherlands	Lesotho	S. Korea	Netherlands
10	Sweden	Belgium	Maldives	Singapore	Belgium
11	India	S. Korea	Solomon Islands	Mexico	Spain
12	Spain	Mexico	Bhutan	Belgium	Australia
13	USA	Taiwan	Comoros	Spain	Singapore
14	China	Singapore	Seychelles	Taiwan	India
15	Norway	Spain	Saint Lucia	Canada	Taiwan
16	Japan	Switzerland	Guinea-Bissau	Arab Emirates	S. Africa
17	Taiwan	Malaysia	Mongolia	Saudi Arabia	Brazil
18	Malaysia	Sweden	Cape Verde	Iraq	Thailand
19	Ireland	Thailand	Grenada	Switzerland	Saudi Arabia
20	Canada	Australia	Fiji	Russia	Canada

each given ranking in 2000, which, given the stability results above, well represents the entire sample period. First note that, apart from ANNS, all "usual suspects" occupy the top-ten positions. Germany scores very high for all statistics but ANNS, while the U.S. and Japan are characterized by a very high rank for weighted statistics but not for ND. This implies that they have relatively less trade partners but the share of trade that they control, the capacity to cluster, and their centrality is very high. Conversely, countries like Switzerland, Italy and Australia have a more diversified portfolio of trade partners with which they maintain less intense trade relationships. Furthermore, it is worth noting that China was already very central in the ITN in 2000, despite its clustering level was relatively lower. India was instead not present among the top-20 countries as far as NS and WCC were concerned; it was only 14th according to centrality and 11th in the ND ranking. Finally, notice how all top-20 countries in the ANNS are micro economies: they typically feature a very low NS and ND, but only tend connect to the hubs of the ITN.

Notwithstanding the presence of a relatively high ranking stability, there are indeed examples of countries moving up or falling behind over the period 1981-2000. For example, as far as centrality is concerned, Russia has steadily fallen in the RWBC ranking from the 6th to the 22th position. A similar downward pattern has been followed by Indonesia (from 17th to 36th). South Africa has instead fallen from 23th (in 1981) to 32th (in 1990) and then has become gradually more central

(16th in 2000). On the contrary, the majority of high-performing Asian economies (HPAE), have been gaining positions in the RWBC ranking. For example, South Korea went from the 24th to the 8th position; Malaysia from the 43th to the 21th; Thailand started from the 42th position in 1981 and managed to become the 18th best central country in 2000. This evidence strongly contrast with the recent experience of Latin American (LATAM) economies (e.g., Mexico and Venezuela) that have – at best – maintained their position in the ranking of centrality (Reyes et al., 2008).

3.4 *Out-of-Sample Evolution*

In the preceding sections, we have investigated the within-sample dynamics of the distributions of node and link statistics. Now we turn out attention to the out-of-sample (long-run) evolution of such distributions by estimating their limiting behavior. To do so, we employ kernel density estimates obtained above to compute ergodic densities, which represent the long-run tendency of the distributions under study[4].

As already noted above, stochastic kernels of all node statistics are quite concentrated and evenly distributed along the 45° line. Therefore, it is no surprising that also their limiting distributions look quite similar to the ones in year 1981. This can be seen in Figure 7, where we exemplify this point by plotting initial vs. estimates of the ergodic distribution for the logs of NS. Both distributions present a similar shape. If any, the ergodic one exhibits a larger variability, a shift to the left of the lower tail and a shift to the right of the upper tail. This can be explained by noticing that the kernel density estimate (Figure 5) shows a relatively larger probability mass under the main diagonal in the bottom-left part of the plot, whereas in the top-right part this mass was shifted above the main diagonal. Such shape-preserving shifts hold also for the other node statistics under analysis. In particular, the ergodic distribution for node RWBC roughly preserves its power-law shape, as well as its scale exponent.

On the contrary, the shape of the stochastic kernel for logged link weights hinted at a concentration of transition densities at the extremes of the range. Middle-range values presented instead a flatter and more dispersed landscape. This partly explains why we observe a radical difference between initial and ergodic distributions of logged link weights. Whereas the initial one is close to a Gaussian (i.e., link weights are well-proxied by a log-normal density), the ergodic distribution displays a power-law shape with very small exponent. This can be seen in Figure 8, where the two plots have been superimposed.

These findings imply that the architecture of the ITN will probably evolve in such a way to undergo a re-organization of link weights (i.e., country total trade volumes)

[4] Given the real-valued statistic X, its ergodic distribution $\phi_\infty(\cdot)$ is implicitly defined for any given τ as $\phi_\infty(x) = \int s^\tau(x|z)\phi_\infty(z)dz$, where $s^\tau(x|z)$ is the stochastic kernel defined in Section 3.2. See also Durlauf and Quah (1999).

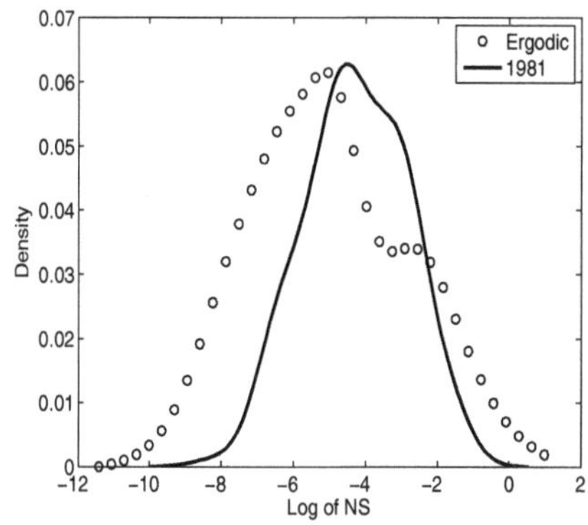

Fig. 7 Node Strength (NS): Kernel density of initial distribution (year=1981) vs. estimate of ergodic (limiting) distribution

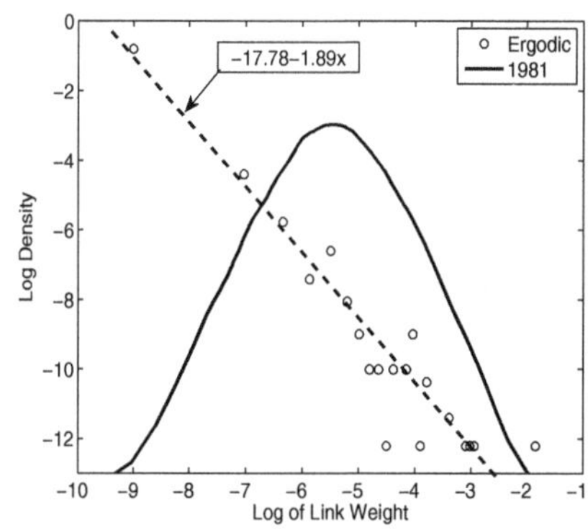

Fig. 8 Positive link weights: Kernel density of initial distribution (year=1981) vs. estimate of ergodic (limiting) distribution. Dotted line: Power-law fit (equation shown in inset). Note: Log scale on y-axis

that is nevertheless able to keep unchanged the most important node topological properties. Such a re-organization seems to imply a polarization of link weights into a large majority of links carrying moderate trade flows and a small bulk of very intense trade linkages. The power-law shape of the ergodic distribution suggests that such a polarization is much more marked than at the beginning of the sample period, when the distribution of link weights was well proxied by a log-normal density. Note, finally, that preliminary exercises indicate that the temporal scale at which this re-organization is expected to occur is of the order of 20-50 time periods (years), provided that the underlying stochastic process would not be hit by structural changes that might undermine its stationarity.

4 Conclusions

In this paper we have explored, from a purely descriptive perspective, the within-sample dynamics and out-of-sample evolution of some key node and link statistic distributions characterizing the topological properties of the web of import-export relationships among world countries (ITN) by employing a weighted-network approach. Our main result is that all the distributions and country rankings of network statistics studied display a rather strong within-sample stationarity. Furthermore, whereas the estimated ergodic distributions of all node-statistics are quite similar to the initial ones, the (positive) link-weight distribution is shifting from a log-normal to a power law. This suggests that a polarization between a large majority of weak-trade links and a minority of very intense-trade ones is gradually emerging in the ITN. Interestingly, such a process is likely to take place without dramatically changing the topological properties of the network.

References

Li, X., Jin, Y.Y., Chen, G.: Physica A: Statistical Mechanics and its Applications 328, 287 (2003)
Serrano, A., Boguñá, M.: Physical Review E 68, 015101(R) (2003)
Garlaschelli, D., Loffredo, M.: Physical Review Letters 93, 188701 (2004)
Garlaschelli, D., Loffredo, M.: Physica A 355, 138 (2005)
Ausloos, M., Lambiotte, R.: Physica A 382, 16 (2006)
Miskiewicz, J., Ausloos, M.: Practical Fruits of Econophysics 5, 312 (2006)
Gorski, A.Z., Drozdz, S., Kwapien, J., Oswiecimka, P.: Acta Physica Polonica B 37, 2897 (2006)
Gligor, M., Ausloos, M.: The European Physical Journal B 63, 533 (2008)
Serrano, A., Boguñá, M., Vespignani, A.: Journal of Economic Interaction and Coordination 2, 111 (2007)
Bhattacharya, K., Mukherjee, G., Sarämaki, J., Kaski, K., Manna, S.: arXiv:0707.4343v1 (2007a)
Bhattacharya, K., Mukherjee, G., Manna, S.: arXiv:0707.4347v1 (2007b)
Garlaschelli, D., Matteo, T.D., Aste, T., Caldarelli, G., Loffredo, M.: The European Physical Journal B 57, 1434 (2007)
Fagiolo, G.: Physical Review E 76, 026107 (2007)
Fagiolo, G., Reyes, J., Schiavo, S.: LEM Working Paper 2007-16, Sant' Anna School of Advanced Studies (2007a)
Tzekina, I., Danthi, K., Rckmore, D.: arXiv:0709.2630v1 (2007)
Fagiolo, G., Schiavo, S., Reyes, J.: Physica A 387, 3868 (2008)
Albert, R., Barabási, A.-L.: Rev. Mod. Phys. 74, 47 (2002)
Dorogovtsev, S., Mendes, J.: Evolution of Networks: From Biological Nets to the Internet and WWW. Oxford University Press, Oxford (2003)
Barrat, A., Barthélemy, M., Pastor-Satorras, R., Vespignani, A.: Proceedings of the National Academy of Sciences 101, 3747 (2004)
Barthélemy, M., Barrat, A., Pastor-Satorras, R., Vespignani, A.: Physica A 346, 34 (2005)
K. Gleditsch, Gleditsch, K.: Journal of Conflict Resolution 46, 712 (2002), `http://privatewww.essex.ac.uk/~ksg/data.html`
Fagiolo, G.: Economics Bulletin 3, 1 (2006), `http://economicsbulletin.vanderbilt.edu/2006/volume3/EB.06Z10134A.pdf`
DeMontis, A., Barthélemy, M., Chessa, A., Vespignani, A.: arXiv:physics/ 0507106v2 (2005)
Saramaki, J., Kivelä, M., Onnela, J., Kaski, K., Kertész, J.: Phsycial Review E 75, 027105 (2007)
Watts, D., Strogatz, S.: Nature 393, 440 (1998)

Newman, M.: Social Networks 27, 39 (2005)
Fisher, E., Vega-Redondo, F.: Working Paper, Cal Poly (2006)
Scott, J.: Social Network Analysis: A Handbook. Sage, London (2000)
Chung, K.L.: Markov Chains with Stationary Transition Probabilities. Springer, Berlin (1960)
Futia, C.: Econometrica 50, 377 (1982)
Spearman, C.: American Journal of Psychology 15, 72101 (1904)
Hollander, M., Wolfe, D.: Nonparametric statistical methods. Wiley, New York (1973)
Braha, D., Bar-Yam, Y.: Complexity 12, 59 (2006)
Reyes, J., Fagiolo, G., Schiavo, S.: LEM Working Paper 2008-10, Sant' Anna School of Advanced Studies (2008)
Durlauf, S.N., Quah, D.T.: In: Taylor, J.B., Woodford, M. (eds.) Handbook of Macroeconomics, vol. 1, ch. 4, pp. 235–308. Elsevier, Amsterdam (1999)

Small World Behavior of the Planetary Active Volcanoes Network: Preliminary Results

Giuseppe Nunnari, Giuseppe Puglisi, Alessandro Bonforte, and Alessandro Spata

Abstract. In this paper we propose a complex network in order to model relationships among active volcanoes of the world. In literature it is well known that volcanoes separated by hundreds of kilometers frequently erupt in unison. These eruptions suggest there is a connection between different active volcanoes. In the attempt to understand and analyze the nature of these relationships we have calculated the correlation among all active volcanoes of the world over the last two thousand years in order to create a functional network. Currently the most relevant result obtained from this study is that the obtained complex network is characterized by small world features.

1 Introduction

Many natural and artificial systems can be modeled as complex networks [1, 2, 3, 7]. Ant colonies, human economies, climate, nervous system, network computers are only a few such examples. One of the reasons for studying such networks is essentially to understand the mechanisms by which information is spread over them. Moreover the discovery of small-world properties of many natural and

Giuseppe Nunnari
Università degli Studi di Catania, viale A. Doria 6, 95125 Catania (Italy)
e-mail: gnunnari@diees.unict.it

Giuseppe Puglisi
Istituto Nazionale di Geofisica e Vulcanologia - Sezione di Catania, Piazza Roma 2, 95125 Catania (Italia)
e-mail: puglisi-g@ct.ingv.it

Alessandro Bonforte
Istituto Nazionale di Geofisica e Vulcanologia - Sezione di Catania, Piazza Roma 2, 95125 Catania (Italia)
e-mail: bonforte@ct.ingv.it

Alessandro Spata
Università degli Studi di Catania, viale A. Doria 6 - 95125 Catania (Italy)
e-mail: alessandro_spata@yahoo.it

R. Menezes et al. (Eds.): Complex Networks: CompleNet 2009, SCI 207, pp. 15–21.
springerlink.com
© Springer-Verlag Berlin Heidelberg 2009

artificial systems [10] has stimulated a great deal of interest in studying the underlying organizing principle of various complex systems. For example in [6] it is shown that several systems such as brains, communication and transportation networks share a common underlying organizing principle, a small-world principle. Recent works have also pointed out the importance of the interconnected networks in geophysics. In [11] the graph theory is applied to model geophysical process such as diffusion and transportation in disordered porous rock. In [9] it is shown that natural fracture/conduit networks can be studied by using a new method based on graph theory. At wider scale, in [5] a graph based approach is applied to analyze seismic activity observed in Southern California. In this work we analyze planetary volcanoes activity from a topological perspective. Although it is well known that volcanoes separated by hundreds of kilometers frequently erupt in unison [8], in this study we apply the network formalism to represent the correlations among active volcanoes with the aim of investigating the possible existence on an underlying organizing principle governing planetary volcanoes activity. We have organized the rest of this paper in the following way: in Sec. II a brief introduction of the small-world theory is given; Sec. III reports the methodology adopted; in Sec. IV results are shown; finally Sec. V draws the conclusions of the paper.

2 Brief Introduction of Small-World Theory

Graphs model the complex information of a system of discrete objects and their relationship. A graph G is defined as a set of N nodes representing the objects and K edges representing the connection among nodes. G is simply represented by its adjacency matrix, an $N*N$ matrix whose entry a_{ij} is 1 if a connection between two generic nodes i, j exists and is 0 if those nodes are unconnected. An important quantity of G is the degree (or connectivity) k_i of a node i defined as the number of edges incident with the node. It is defined in terms of the adjacency matrix as:

$$k_i \equiv \sum_{j=1}^{N} a_{ij} \tag{1}$$

The average value of k_i is $k = 2K/N$. The structural properties of a graph are usually quantified by two quantities: the characteristic path length L and the clustering coefficient C [10]. L is the average distance between two generic nodes and is defined as follows:

$$L = \frac{1}{N(N-1)} \sum_{i \neq j} d_{ij} \tag{2}$$

where d_{ij} is the shortest path lengths between two generic node i,j. C is a local property of the graph and represent the average fraction of pairs of neighbours of a node which are also neighbours of each others. It is defined as:

$$C = \frac{1}{N} \sum_{i} C_i \tag{3}$$

where C_i is the number of edges existing in the subgraph G_i of the neighbours of i, divided by the maximum possible number $k_i(k_i-1)/2$. The study we present here is strictly related to three kinds of graph topologies: regular, random and small-world. A regular graph is a graph whose nodes all have the same number of connections and are connected to their immediate neighbours. Regular graphs have a high clustering coefficient $C = 3/4(k-2)/(k-1)$ but a long characteristic path length $L \sim N/2k$. A random graph is a graph in which connections among nodes are determined in some random way. Random graphs have a low clustering ceofficient $C = k/N$ but the shortest possible path length $L \sim ln(N)/ln(k-1)$. Small-world graphs [10] are a class of network whose topology is intermediate between regularity and randomness. These networks combine the high clustering of regular graphs and the short characteristic path lengths of random graphs. Very interesting is the fact that many real world networks, both natural and artificial, are small-world networks, i.e. efficient systems that are able to quickly transfer any quantity.

3 Method

In order to achieve a graph representation of relationships among active volcanoes we use a methodology that resembles the approach proposed in [5]. At each time step the activity function $a(t)$ of a generic volcano is defined in terms of a binary time-series as follows: "1" if the volcano is erupting, "0" otherwise. Evaluating for each pair of volcanoes v_1 and v_2 the Pearson correlation coefficient given by

$$r(v_1, v_2) = \frac{\sigma_{v_i v_j}}{\sigma_{v_i} \sigma_{v_i}} \tag{4}$$

where $\sigma_{v_i v_j}$ is the covariance between the activity functions of the volcanoes v_1 and v_2 and σ_{v_i} and σ_{v_j} are their standard deviations, we obtain the symmetric square matrix $Corr$, referred to as the correlation matrix, consisting of all correlation values among active volcanoes. In order to represent these relationships as a graph we have to define both its nodes and edges. The volcanoes are the nodes of the graph. In order to establish the edges we fix a positive threshold value Th and define the adjacency matrix M as follows:

$$M_{ij} = \begin{cases} \text{"0" if } Corr_{ij} \geq T_h \\ \text{"1" if } Corr_{ij} < T_h \end{cases} \tag{5}$$

In this manner when the correlation between two volcanoes activity is higher than the threshold value the respective nodes of the graph are connected by an edge.

4 Results

In order to perform our analysis we have examined global volcanic activity information available for the past two thousand years. We have used the Smithsonian Institution, Global Volcanism Program Catalogue of volcanic eruptions. A suitable software tool and a relational database were implemented in order to systematically perform graph analysis. All activity functions $a(t)$ were calculated assuming a time resolution t_r of 1 year. As noted in [9] the threshold Th plays a key role in the connectivity of the generated graph. In this study we have experimentally found that for Th value greater than 0.6 the graph is disconnected. In this case the generated network consists of several disjoint clusters of few high correlated volcanoes (volcanoes that erupt in unison) connected by strong links. Therefore, due to the disconnected nature these graph are not suitable to study relationships among active volcanoes on a planetary scale. By decreasing the threshold value a lot of weaker links among volcanoes are obtained.. The terms strong and weak links are used according to the Granovetter work [4] . For these reasons here we analyze the structural properties of the connected components of the generated graph by using Th values lower than 0.6. In this way the isolated nodes or the fragmented small clusters of the resulting networks are not taken into account. This choice also avoids dealing with the infinite distance among nodes belonging to different disjoint clusters.

Before showing the complete graph analysis we emphasize some interesting correlations obtained on a volcanoes network generated by using a threshold value $Th = 0.5$. A first-order correlation among the volcanoes surrounding the Pacific plate is immediately evident, suggesting a common tectonics driving the volcanic activity all around the Pacific, even for volcanoes that are rather distant from each other. In fact, looking for example at the volcanoes forming the Aleutian island arc (Fig .1), that marks the convergent boundary between north-American and Pacific plates, a very interesting correlation is revealed, not only among them (especially among the Alaskan ones) but also between these and other volcanoes lying all around the pacific plate. It is also interesting to note also a significant correlation among the South America's volcanoes (Fig. 2), which suggests that the collisional tectonics between the Pacific and South America plates feeding the volcanic activity is fairly uniform at least along the entire Chilean coast.

In Fig. 3 a more complex correlation affecting Indonesian volcanoes is also shown; in this area, volcanoes lie on the border of a sort of *peninsula* on the south-easternmost part of the Eurasia plate, facing the Australian one. Here, volcanoes are not well correlated among them and show few correlations with other volcanoes lying on the opposite side of this *peninsula*, facing the Philippine and Pacific plates. On the contrary, they show more correlations with other volcanoes outside the SE-Asia plate, especially with those lying in the northern and north-western Pacific. This feature, and eventually others that will become evident during a deeper investigation, may suggest more complex interrelationships between different tectonic plates and needs to be further analyzed.

We have compared the clustering coefficient C and the characteristic path length L of the connected components of the analyzed volcanoes network with the corresponding theoretical values computed for both a random graph and a regular graph with the same number of nodes and the same average node degree. The aim of this comparison is to characterize the behavior of our network. It should be kept in mind that small world networks are characterized by high C values and small L.

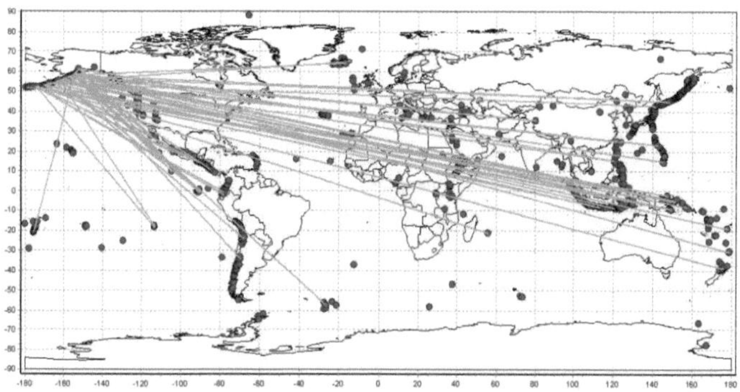

Fig. 1 It is interesting to note a correlation of the North Pacific volcanoes almost exclusively with other volcanoes around the Pacific plate ($Th = 0.5$)

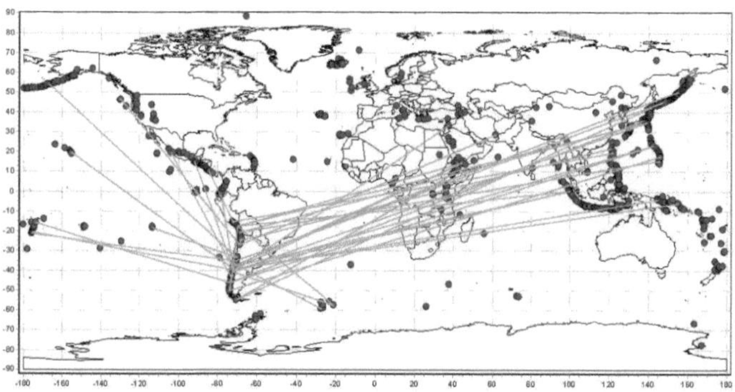

Fig. 2 Again, South Americas volcanoes show correlation among them and mainly with the opposite side of the Pacific ($Th = 0.5$)

The theoretical and the computed values for Th values ranging between 0.0 to 0.55 are shown in Fig. 4. Since volcanoes networks exhibit clustering coefficients that are always much greater than those of the random graphs and average path lengths slightly higher than those of random graphs, it is evident that these networks are characterized by marked small world features.

These results have been found considering a time resolution t_r of one year, i.e two volcano eruptions happen *at the same time* if they occur in the same year. The matrix correlation, obviously, depends on this choice. We have repeated the above analysis using different values of t_r and we have found that increasing its value the quantity L tends to decrease and C tends to increase. This was expected because, obviously, the higher t_r the greater are both the active volcanoes and, consequently, the links among them. Furthermore, in the limit $t_r \rightarrow \infty$ the graph tends to be "all to all" connected thus explaining both L and C trends.

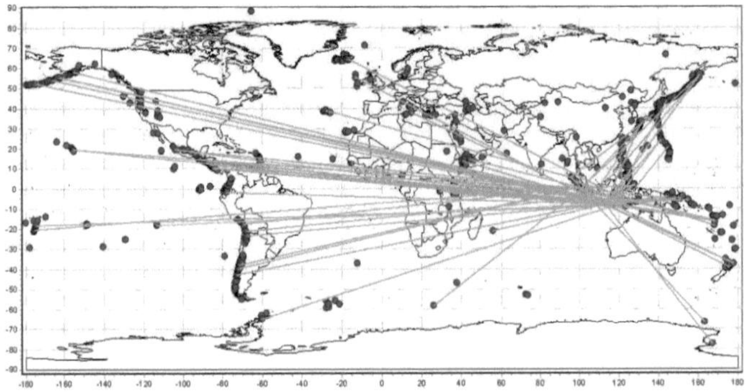

Fig. 3 A more complex correlation affects SE Asia volcanoes, showing connections mainly with western and northern Pacific volcanoes but also with many others around the world ($Th = 0.5$)

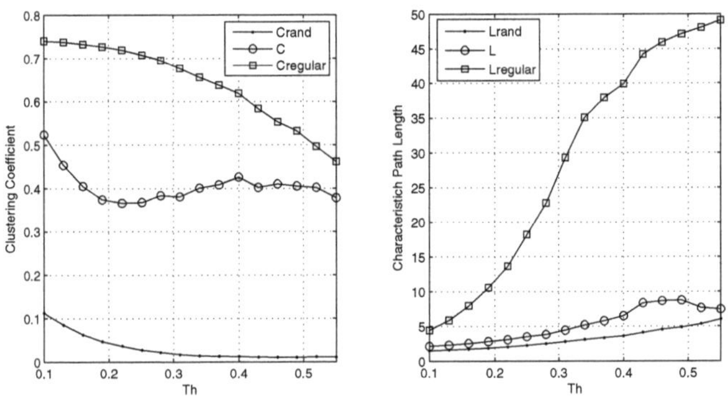

Fig. 4 Clustering coefficient (left) and characteristic path length (right) of the networks for Th values ranging from 0.1 to 0.55

5 Conclusions

We propose a complex networks based approach in order to model and characterize volcanoes activity correlation observed on a planetary scale over the last two thousand years. Worldwide volcanic activity is in fact related to the general plate tectonics that locally drives the faults activity, that in turn controls the magma upraise beneath the volcanoes. To find correlations among different volcanoes could indicate a common underlying mechanism driving their activity and could help us interpreting the deeper common dynamics controlling their unrest. The understanding of eventually detected correlations could be also used to further implement warning systems, relating the unrest probabilities of a specific volcano also to the ongoing activity to the correlated ones. Our preliminary results suggest that, as for other many physical and biological systems, an underlying organizing principle of planetary volcanoes activity might exist and it could be a small-world principle.

In fact we found that, from a topological perspective, volcanoes correlations are characterized by the typical features of small-world network: a high clustering coefficient and a low characteristic path length. These features confirm that global volcanoes activity is characterized by both short and long-range correlations. We stress here the fact that numerical simulation carried out in this work seems to agree with geological evidences (eg. the Pacific plate, South America volcanoes activity and so on). However a detailed analysis of numerical correlation pointed out in this work and geological implication requires a lot of effort and is still running. Thus this work represents preliminary contribution to better understand and clarify, from a geophysical point of view, the nature of planetary correlations among active volcanoes. Further work is needed.

References

1. Albert, R., Barabasi, A.-L.: Rev. Mod. Phys. 74, 47 (2002)
2. Bar-Yam, Y.: Dynamics of Complex System. Addison-Wesley, Reading (1997)
3. Boccaletti, S., Latora, V., Moreno, Y., Chavez, M., Hwang, D.-U.: Phys. Rep. 424, 175 (2006)
4. Granovetter, M.S.: The strength of weak ties. American Journal of Sociology 78(6), 1360–1380 (1973)
5. Jiménez, A., Tiampo, K.F., Posadas, A.M.: Small world in a seismic network: the California case. Nonlin. Processes Geophys. 15, 389–395 (2008)
6. Latora, V., Marchiori, M.: Efficient behavior of small-world network. Phys. Rev. Lett. 87, 1987011–1987014 (2001)
7. Newman, M.E.J.: SIAM Rev. 45, 167 (2003)
8. Simkin, T., Siebert, L.: Volcanoes of the World. Geoscience, Tucson (1994)
9. Valentini, L., Perugini, D., Poli, G.: The 'Small-World' Nature of Fracture/Conduit Networks: Possible Implications for Disequilibrium Transport of Magmas Beneath Mid-Ocean Ridges. Journal of Volcanology and Geothermal Research 159(4), 355–365 (2007)
10. Watts, D.J., Strogatz, S.H.: Nature (London) 393, 440 (1998)
11. Yang, X.-S.: Small-World Networks in Geophysics. Geophys. Res. Lett. 28(13), 2549–2552 (2001)

Correlation Patterns in Gene Expressions along the Cell Cycle of Yeast

Jelena Živković, Marija Mitrović, and Bosiljka Tadić

Abstract. Currently available genome-wide expression measurements and the respective data bases represent the reliable entry information for gene interaction research. In order to unravel the collective behavior of genes, methods and approaches pertinent to complex dynamical systems are necessary. Using the network theory, we study correlation patterns in the time series of gene expressions of Yeast measured along the cell cycle. We select a subset of genes by their leading participation in the scale-invariant features of the expression data. Applying standard filtering of the correlation matrix reveals inhomogeneous mesoscopic structure of the related graph with several well defined modules of genes. The findings are corroborated by the spectral analysis of the correlation matrix and the eigenvector localization on the graph. The topologically distinct groups of genes which are co-expressed within a given phase of the cell cycle belong to different functional categories but often share the same localization, i.e., nucleus, cytoplasm, or mitochondria, inside the cell.

1 Introduction

Mapping of a complex dynamical system on a mathematical graph, i.e., by identifying its nodes and their connections (edges), provides a ground for quantitative study of complexity by methods of the formal graph theory and statistical physics of structured networks [1, 2]. In this picture, the dynamic processes of the system

Jelena Živković
SPM Group, IMM, Faculty of Science, Radboud University, Nijmegen, The Netherlands
e-mail: jelena.zivkovic@gmail.com

Marija Mitrović
Scientific Computing Laboratory, Institute of Physics, Belgrade, Serbia
e-mail: mitrovic@scl.rs

Bosiljka Tadić
Department of theoretical physics, Jožef Stefan Institute, Ljubljana, Slovenia
e-mail: bosiljka.tadic@ijs.si

R. Menezes et al. (Eds.): Complex Networks: CompleNet 2009, SCI 207, pp. 23–34.
springerlink.com © Springer-Verlag Berlin Heidelberg 2009

can be viewed as a set of *time series* related to the fluctuations of the activity of each node on the graph. Such time series are often measured in the real systems and massive empirical data are currently available, for instance, the time series in the stock-price fluctuations in the stock market [3] and temporal fluctuations of the activity of routers in the Internet traffic [4]. Analysis of the time series and models reveals the occurrence of the long-range correlations and collective behavior related to the structure of the underlying dynamical system [4, 5, 6].

Studies of the correlations in time-series were attempted using various methods in order to unravel the underlying functional structure of the complex system, which results in these time series. Owing to the nonlinearity of the dynamics and the complexity (emergence) of the *collective dynamical effects* which do not occur at the level of isolated units, this problem represents a major challenge in contemporary science of complex systems. Often the mathematically elaborate methods theoretical models are necessary to interpolate between the observed data and true interactions in the system. On the other hand, massive data of genome-wide expressions measurements [7] are available, which contain hidden information about gene interactions. Assuming a model of gene dynamics, analysis of gene expression data yielded rather limited information about *pairwise* gene interactions (gene network) [8, 9]. Using formal analysis we recently studied the traffic time series on known modular networks [10]. We have shown that, unlike individual links, the network modularity (structure at the mesoscopic scale) can be revealed quite accurately from the correlations in the time series. The accuracy increases when the network is fully partitionable into modules. (See also Refs. [11] and [12] for different approaches to the problem of resolution of the network community structure.)

In this work we study correlations in the empirical time series of gene expressions measured [13] for each gene in the genome of Yeast *Saccharomyces cerevisiae (S.c.)*. Compared to the typical time series, say in the stock prices or Internet traffic, the time series of the gene expressions differ in the following: the expressions are sparsely measured (every 10 minutes) resulting in fewer number of points (the statistical importance of the data points is discussed in [13]); more importantly, the time evolution is naturally related to the *cell cycle*, therefore, when a cycle is completed (approximately 2 hours), the gene activity returns to the beginning of a new cycle. Our focus here is on the gene clustering that can be detected through the analysis of these time series. Our analysis is based entirely on the expression data without further reference to the transcription regulation between genes with already known transcription factors [7, 2]. We also restrict the analysis to a subset of 603 genes whose activity is most prominent in one of the phases of the cell-cycle (the "cell-cycle type"). Identification of the cell-cycle related genes and their mutual correlations is an important and still open problem (see [14] and references therein).

In Section 2 we present the statistical analysis of the genome-wide expression data and select the genes with leading contributions in the cell-cycle phases. We further analyze the correlations of the selected subset of genes in Section 3, where we determine the groups of genes using filtered correlation matrix and its eigenvalue spectrum. Section 4 is devoted to the study of topology of the network constructed on the basis of correlations, and Section 5 makes a brief summary.

2 Scale Invariance in the Expressions of CC Genes

We consider empirical data for the time-course expression of the whole genome of yeast *S.c.*, measured in Ref. [13] at 17 equidistant time points along two full cell cycles. The statistical analysis of the data [15] (see also [2]) revealed scale-invariance in the ranking of the gene expressions and the broad distributions with power-law tails, which are characteristic for collective behavior, e.g., in self-organized dynamical systems. In particular, the average expressions of genes during the cell cycle obey a power-law ranking, also known as Zipf's statistics [21, 22], which indicates different contributions of genes to the dynamics during the cell cycle. Here we use the scale-invariance of the expression data to select the genes which contribute at most to the observed scale-invariance.

For this purpose we analyse the *differential expression* which is defined as $\Delta X_i(t) = h_i(t) - h_i(t-1)$ for each gene $i = 1, \cdots, N = 6406$ ORF (genes). The ranking distributions of all measured entries ($N \times 17$) is given in Fig. 1a, top curve. Similar ranking statistics for all N genes but for a single time at $t = 1, 8, 16$ are also given by other three curves. Broad distribution with the Zipf's law suggests *non-random correlations* for genes with the expressions above a threshold (marked by the horizontal line, which actually coincides with often used discrimination level $2 \times h_0$, where h_0 is the average expression in the whole system and for all instances of time). The occurrence of a smaller slope for genes with the largest expressions (upper part of the ranking curve) suggests that a community structure might be present with several genes playing a central role in each community. The scale-invariance of the expressions is also seen in the distribution for the differential expressions in Fig. 1b, again for the three time points and for all data. This distribution exhibits a power-law tail after a characteristic scale $\sim \Delta_0$, and can be fitted by the q-exponential form

$$P(\Delta X) = B_q \left[1 - (1-q)\frac{\Delta X}{\Delta_0}\right]^{\frac{1}{1-q}} \; ; \; q \neq 1 \quad , \tag{1}$$

where $q \simeq 1.36$ represents the non-extensivity parameter [20].

For the purpose of this work we select the genes with the expressions above the threshold $< h_i >> 2h_0$, as explained above and marked in Fig. 1a. The selected set consists of $N_s = 1216$ genes. Among these the first 612 in the ranking order are the genes which are expressed throughout the entire cell cycle, while the following 604 genes are mostly expressed within one or two of the cell cycle phases (G1,S,G2,M). In the following we will focus on this latter group of genes (cell-cycle genes). For the illustration, in Fig. 2 are shown the expressions for several cell-cycle genes within the first cycle.

For each pair of genes we compute the correlation coefficients of the temporal differential gene expressions $\Delta X_i(t)$, which is given by

$$C_{ij}(t-t') = \frac{\Sigma_t(\Delta X_i(t) - \langle \Delta X_i \rangle)(\Delta X_j(t-t') - \langle \Delta X_j \rangle)}{\sigma_i \sigma_j} \quad . \tag{2}$$

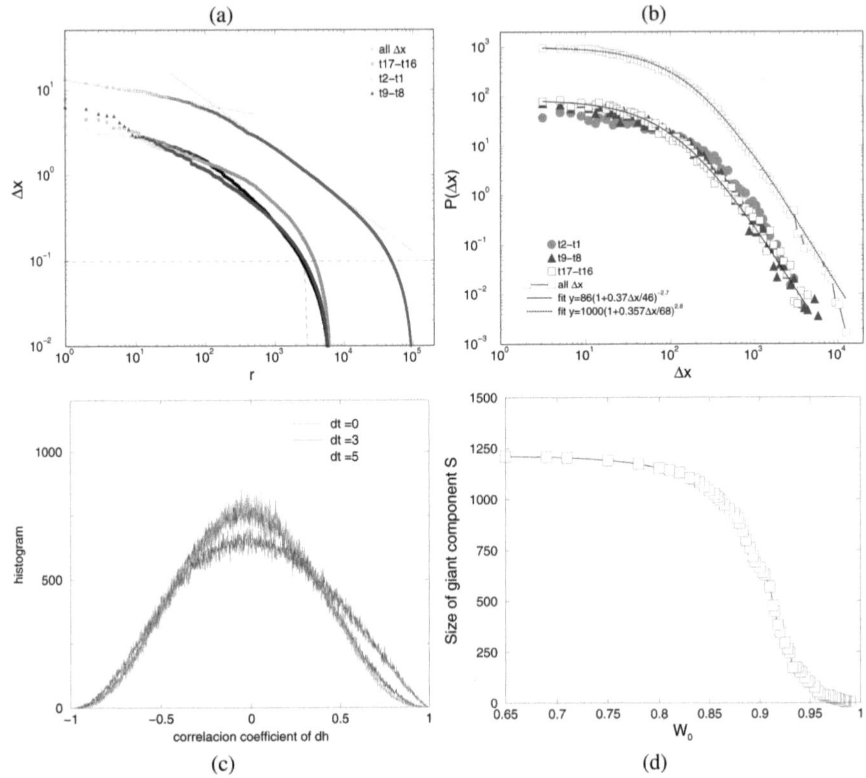

Fig. 1 (a) and (b) Statistical analysis of the differential expressions data for all genes in the genome of yeast *S.c.* shown in ranking order (a) and histogram (b). In both panels, top curve includes data of all measurements, while lower three curves are for all genes at a given instant of time in the cell cycle. The discrimination dashed line in (a) indicates how the first $N_s = 1215$ genes in the ranking order are selected. (c) For the selected genes: the distribution of correlation coefficients C_{ij} for co-expression ($dt = 0$) and for two time-delayed correlations. (d) Size of the giant component as a function of the threshold correlation W_0 in the selected set of N_s genes

where σ_i and σ_j are the standard deviations of the respective time series of the genes i and j. The distribution of the correlation coefficients for the selected set of N_s genes is given in Fig. 1c, where $dt \equiv t - t'$ indicates the time-lagged correlations. As the Fig. 1c shows, in this set of genes the strongest deviations from normal distribution are seen in the equal-time correlations $dt = 0$, which we will consider in the following. In order to extract the relevant correlations, e.g., in the tail of the distributions in Fig. 1c, from the random correlations around zero, one considers only the correlations above a threshold value $|C_{ij}| > W_0$. However, for the analysis of the community structure, the threshold needs to be low enough such that all nodes of the network based on C_{ij} as the connectivity matrix belong to a single connected component. Large values of the threshold lead to fragmentation of such network. On the other hand, we may observe how the giant cluster is formed by slowly reducing

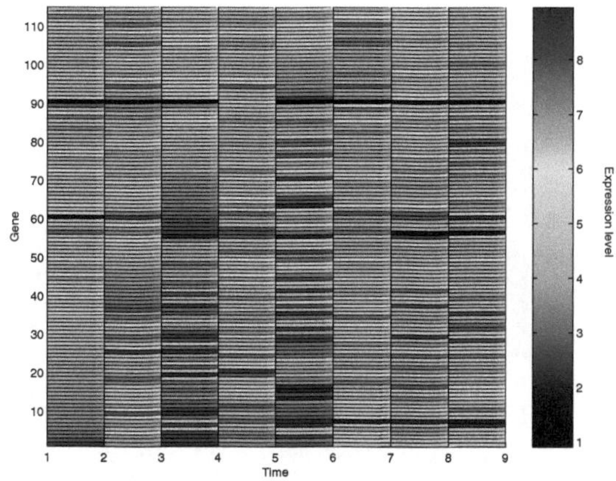

Fig. 2 Three-dimensional plot of the gene expressions within the first cycle for several leading Cell-Cycle genes ordered according the time of their maximum expression

the threshold starting from $W_0 = 1$ (in Ref. [15] formation of the giant cluster for the entire genome of Yeast was studied). In the case of selected N_s genes which we consider here, the largest cluster S_{max} as function of the threshold W_0 forms slowly and all N_s genes are connected only at $W_0 \geq 0.6$, as shown in Fig. 1d. Therefore, we will consider the critical threshold $W_0 = 0.6$ for the study of the correlation matrix of N_s genes and their subset—CC genes, in the next session.

3 Detecting Modules in the Correlation Matrix

As mentioned above, the network based on the correlation matrix **C** with the elements $C_{ij} > W_0 = 0.6$ is connected, however, it has very large number of links. As mentioned in the Introduction, in order to find a meaningful structure in the correlations, one needs to apply a filtering procedure in order to reduce the number of 'spurious' links. One of the methods which we use here is based on affinity transformation [17]. For the filtering procedure[17], the matrix elements C_{ij} of the correlation matrix are first mapped to the positive interval $[0,1]$. Then each element C_{ij} is multiplied with a factor M_{ij} which is constructed from the elements of the rows i and j in the correlation matrix: Excluding the diagonal elements C_{ii} and C_{jj}, the remaining matrix elements are first reordered to form the $n \equiv (N-1)$-dimensional vectors $\{C_{ij}, C_{i1}, ..., C_{iN}\}$ and $\{C_{ji}, C_{j1}, ..., C_{jN}\}$. Then with the components of these vectors the Pearson's coefficient M_{ij} is computed according to the general expression in Eq. (2). The matrix elements C^M_{ij} of the *filtered correlation matrix* $\mathbf{C^M}$ are given by the respective products

$$C^M_{ij} = M_{ij} C_{ij} \,. \tag{3}$$

In this way, the correlation between the nodes i and j is enhanced if the corresponding meta-correlation element M_{ij} is large, i.e., the nodes i and j are connected to the rest of the system in a similar way (which will be also manifested in their time series), and reduced otherwise. Therefore, after the filtering procedure we expect to

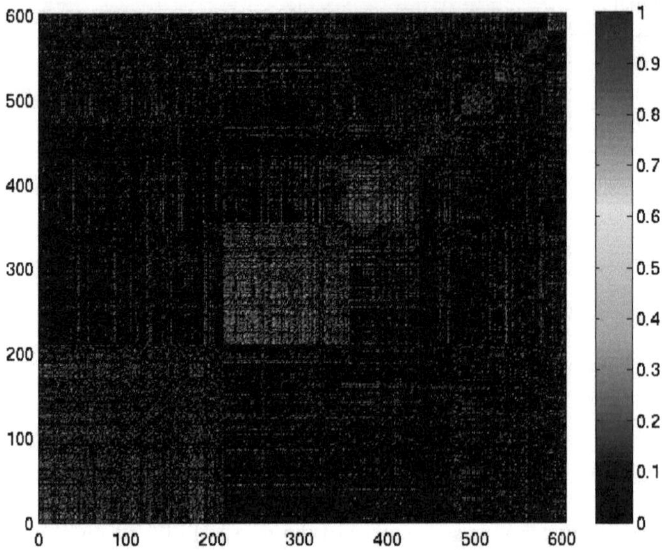

Fig. 3 Filtered correlation matrix showing several modules (diagonal blocks) of CC genes

find enhanced correlations between 'similar' nodes, and reduced in the case when nodes play different role in the system. Applying then the same threshold as above, we find groups of nodes with enhanced correlations inside the group and generally reduced correlations (number of links above the threshold) outside the group. In

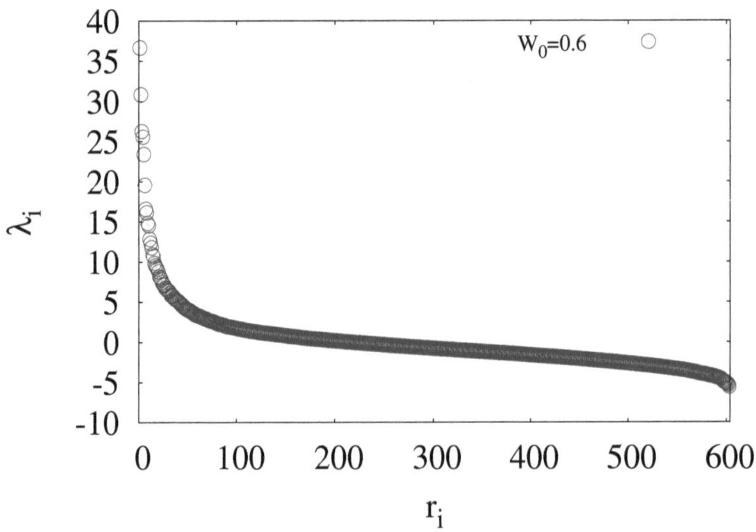

Fig. 4 Eigenvalues of the filtered correlation matrix of CC genes ranked according to value

Fig. 3 we show the filtered correlation matrix of the 603 CC genes. The filtered matrix clearly exhibits the modular structure with several modules, blocks along the diagonal, of different sizes. It is also clear that some of the modules are not entirely homogeneous, having a strongly expressed gene inside the module leads to enhanced correlations between the other genes in the module and with other modules. As shown in ref. [10] for the model network, the internal inhomogeneity of the modules leads to extra correlations between the modules, which can not be filtered out by standard methods. However, the size and the association of the nodes with the modules is correctly matched. We expect the same is true in the case of gene correlations. The structure of the modules can be visualized and analyzed in detail using the network representation, with the adjacency matrix defined by the elements of the filtered correlation matrix $\mathbf{C^M}$. In the following we first analyze the spectrum and subsequently the structure of the network.

3.1 Spectral Analysis of the Gene Correlation Matrix

The modular structure of the network can be also visualized via the spectral analysis of its connectivity matrix and other matrices, e.g., Laplacian, related to the structure [16]. Here we solve the complete eigenvalue problem of the filtered correlation matrix shown in Fig. 3. The matrix with weighted links (excluding the self correlations) above the threshold 0.6 is considered. The results for the eigenvalues λ_i are shown in Fig. 4, ranked according to their values. In the eigenvalue spectra of structured networks [23] the largest eigenvalue is separated from the rest of the

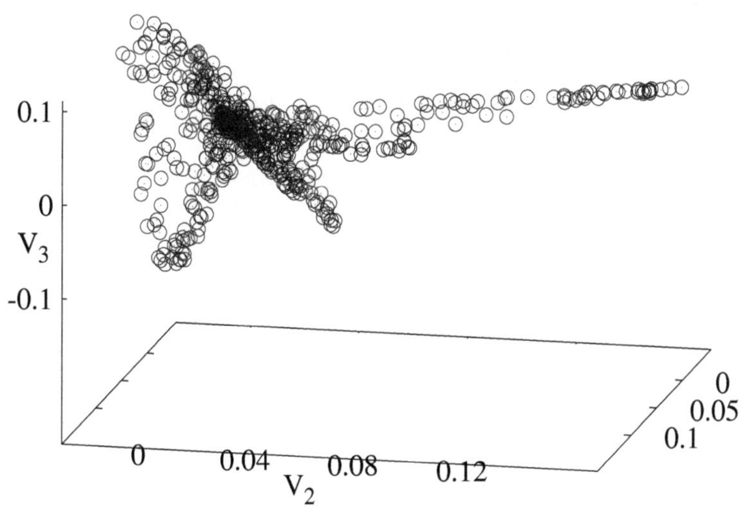

Fig. 5 Scatter plot: Components of the eigenvectors associated to three large eigenvalues $\lambda_3 = 25.568807$, $\lambda_2 = 26.276988$, and $\lambda_1 = 30.840979$. Most distant points on the plot correspond to the genes DIM1, CLB5, and TY2B, located in nucleus, cytoplasm, and mitochondria, respectively

Table 1 Characteristics of genes associated with four tips of the branches in the scatter plot in Fig. 5. Data according to MIPS database [19]. Abbreviations: UP (UNCLASSIFIED PROTEINS); BIOG (BIOGENESIS), M (METABOLISM), CC (CELL CYCLE), CT (CELLULAR TRANSPORT), E (ENERGY), CTD (CELL TYPE DIFFERENTIATION), TR (TRANSCRIPTION), ENV (Interaction with Environment), PF (regulation of PROTEIN FUNCTION), PM (PROTEIN FATE , PROTEIN MODIFICATION), PS (PROTEIN SYNTHESIS), DEV (DEVELOPMENT), CF (CELL FATE), CR (CELL RESCUE); Localization: CP (cell periphery), CYT (cytoplasm), NUC (nucleus), VAC (vacuole), Ec (extracellular), MIT (mitohondria), ER (endoplas.reticulum)

ORF reading	Gene Name	Functional Category	Disruption	Localization
YPL266W	DIM1	TRANSCRIPTION	letal	NUC
YOL005c	RPB11	DNA binding, TR	lethal	NUC
YPL267W	ACM1	UP	viable	NUC
YKR013W	PRY2	UP	viable	NUC, VAC
YNL281w	HCH1	CELL RESCUE	viable	CYT,NUC
YDL039c	PRM7	DEVELOPMENT	viable	?
YDL003W	MCD1	CC & DNA processing	lethal	CYT, NUC
YAL036c	RBG1	GTP binding	viable	CYT
YJL174W	KRE9	CTD, M, BIOG, E	lethal	Extracellular
YDL241W	-	UP	viable	CYT,NUC
YPL235W	RVB2	M, CC & DNA processing, TR	lethal	?
YBL101W-b	TY2B	Viral& PP	-	(cyt)
YLR257W	-	UP	viable	CYT
YFL035c-b	MOB2	CC & DNA, PM, M, PF	lethal	CP, CYT
YHR135C	YCK1	ENV, CC & DNA, PM, M, CTD	viable	PM, CYT, ER, NUC
YPL226W	NEW1	M, PB, PS	viable	CYT,NUC,MIT
YML085C	TUB1	DEV, CC & DNA processing, M	lethal	CYT, cytoskeleton
YKL001C	MET14	METABOLISM	viable	CYT
YDR245W	MNN10	CTD, M, BIOG, PM	viabl	ER, golgi
YPL032C	SVL3	CTD, CF, BIOG	viable	CP, CYT
YGR118W	RPS23A	PROTEIN SYNTHESIS	viable	CYT
YPL028W	ERG10	METABOLISM	lethal	CYT, NUC
YDR346c	-	unclassified	viable	CYT, NUC
YDL124w	-	PB, M	viable	CP, CYT, NUC
YBL064c	PRX1	CR, DEFENSE & VIRULENCE	viable	CYT, NUC, MIT
YJL217w	-	UP	viable	CYT
YBR194w	SOY1	UP	viable	CYT, NUC
YER001w	MNN1	M, PM	viable	golgi, VAC
YNR065c	YSN1	UP	viable	?
YDL166c	FAP7	CR, TRANSCRIPTION	letal	CYT, ER, NUC
YEL003w	GIM4	BIOG, PM, PB	viable	CYT
YDR098c	GRX3	CR, PM, CT	viable	CYT, NUC
YAR002c-a	ERP1	CT, PM	viable	golgi, MIT, VAC
YKL142W	MRP8-	PROTEIN SYNTHESIS	viable	CYT, MIT
YPR120C	CLB5	CC& DNA processing	viable	MIT
YKL190w	CNB1	TR, PB	viable	CYT
YPL135w	SU1	M, ENV	viable	MIT
YCR070w	CPR4	PROTEIN FATE MODIFICATION	viable	ER, VAC
YLR017w	MEU1	METABOLISM	viable	CYT, NUC
YML010W	SPT5	TR, CC&DNA processing	viable	NUC, MIT
YER074w	RPS24A	PROTEIN SYNTHESIS	viable	CYT, MIT
YBR048w	RPS11B	PROTEIN SYNTHESIS	viable	CYT
YOR153w	PDR5	ENV, CT, PB, CR	viable	CP, CYT, MIT

spectrum and the components of its eigenvector are related to the eigenvector-centrality measure [1, 16]. In the presence of the structural modules (or communities), additional eigenvalues appear between the largest eigenvalue and the main part of the spectrum. The number of such eigenvalues is directly related to the number of topologically distinct modules. In the upper part of the plot in Fig. 4 six such eigenvalues occur, corresponding to six (sufficiently large and distinct) modules in the correlation matrix.

Further remarkable property of the eigenvalue problem of a modular network is illustrated by the localization of the eigenvectors of the large eigenvalues of its connectivity matrix (or similarly, of the small non-zero eigenvalues of the Laplacian matrix [16]) on the network modules. Formally, the term *localization* of the eigenvector $V_i \equiv V(\lambda_i)$ belonging to the eigenvalue λ_i denotes that it has a *nonzero component* V_i^κ, corresponding to the node with the index κ. In modular networks the nodes with indexes κ corresponding to positive/negative values of the eigenvector components belong to different modules. Moreover, the extended branched view of the scatter plot indicates a structured network. In our case with the filtered correlation matrix we show the eigenvectors associated with the three large eigenvalues just below the λ_{max}. The 3-dimensional scatter plot of the components these eigenvectors $(V_1^\kappa, V_2^\kappa, V_3^\kappa)$ is shown in Fig. 5. Each point in this plot correspond to the same index κ of the three eigenvectors, and thus represents one node on the network. The separate branches in the scatter plot correspond to different modules on the network (diagonal blocks in its adjacency matrix). In our case each node represents a specific gene. Therefore, genes belonging to different branches can be identified by their names and other known biological properties [19]. For the illustration, the genes at the far ends in the scatter plot are node with the index (cf. matrix in Fig. 3) 471 corresponding to gene DIM1 or ORF "YPL266W"(most right in Fig. 5), index 130 corresponding to the gene CLB5 or "YPR120C" (at the front tip of the middle branch), index 367 or "YOR153W" and gene PDR5 (lower left tip), and index 221 representing the gene TY2B "YBL101W-b" (top of the middle branch in the scatter plot on Fig. 5). Identity of more genes grouped in the vicinity of these four are listed in the separate sections in the Table 1. Their biological properties, taken from the MIPS database [19], indicate that the genes in each group span the whole spectrum of different biological functions, however, they mostly share the same physical localization in the cell (nucleus, cytoplasm, mitochondria, cell periphery, etc.).

4 Structure of Correlation Network and Group-Identity of Genes

The structure of the modules of co-expressed genes can be studied systematically by analysis of topology of the network whose adjacency matrix is based on the filtered correlation matrix. In Fig. 6a we show such network with nodes representing genes and edges indicating the expression correlations above the threshold (shown are only the links $C_{ij} > 0.8$). Colors (red, green, yellow, blue) on the nodes represent one of the phases (G1, S, G2, M) of the cell cycle where the gene has its peak expression. The presence of modules are seen on the network, however, the number of links inside these modules is too large. Therefore, we use the *spanning tree* as another way to visualize the structure of these weighted networks. We construct the spanning tree of the gene correlation network (shown in Fig. 6b), where each gene is connected to the rest of the tree by its *strongest* link (maximum correlation spanning tree). Separate branches on the tree correspond to seven modules in the correlation matrix (cf. Fig. 3). The predominant color in parts of the tree (branches) indicates gene groupings according to the phase of their peak activity. Detailed identification of the genes within a branch or sub-branch, an example is displayed in Fig. 6c with the names of genes, shows that genes within the same phase of the cell cycle have

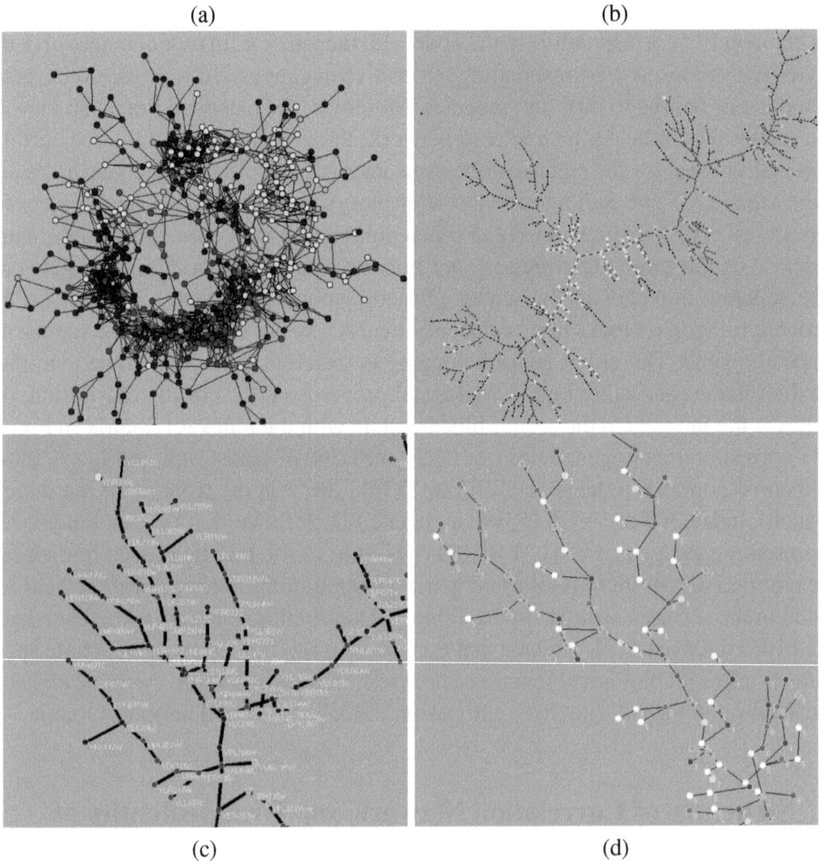

Fig. 6 (a) Network based on the filtered correlation matrix of CC genes (shown are links $C_{ij} > 0.8$)). (b) Maximum correlation spanning tree of the network. (c) Zoomed branch of the tree with genes marked by their ORF labels in the original data. Colors on nodes in (a-c) indicate the cell-cycle phases (G1:red, S:green, G2:yellow, M:blue) in which the gene has maximum measured expression. (d) Biological functions (grouped in seven categories) for genes within same branch

variety of biological functions. An example of the tree branch is shown in Fig 6d, with gene functions indicated by colors and numbers from 0-7 (for seven groups of known gene functions: metabolism, energy, cell cycle/cell rescue, transcription, protein synthesis, cellular transport, cell type differentiation and development, and unknown). The genes along the same branch of the tree have additional similarity in their localization in the cell. For instance, according to the MIPS database [19], the majority of the genes belonging to the branch in Fig. 6c are localized either in the nucleus or in the nucleus and cytoplasm.

5 Conclusions

We have studied correlations in the activity of genes of yeaast *S.c.* based on the empirical data [13] in the form of time series, in which the fluctuation in the expression of each gene are measured along the cell cycle. We employed three methods suitable for the analysis of *collective dynamical behavior* in self-organized systems:

- *Scale-invariance* in various statistical measures of the differential expressions suggests uneven role of different genes in the cell-cycle dynamics. With the ranking distribution (Zipf's law) we selected the genes with leading contributions to the scaling behavior, what implies prominent correlations within the selected set.
- *Spectral analysis* of the appropriately filtered correlation matrix is suitable for detecting modules of strongly correlated genes. We find several such modules of different size in the subset of the cell-cycle genes. The eigenvectors associated with the largest eigenvalues of the correlation matrix show a pattern of localization of the non-zero components on gene indexes belonging to different modules.
- *Network topology* based on the filtered correlation matrix (above a threshold) gives a systematic survey of the content of different groups (modules). Identifying the genes in different modules suggest that they share similarity in the phase of the cell cycle with their peak activity and also in the physical localization inside the cell, checked against the biological information in MIPS database.

We have shown that the genes co-expressed along the cell-cycle show certain patterns of correlated activity which is unraveled by the formal analysis within the network theory without further reference to bio-chemical interactions (transcription regulations) or a specific mathematical model for the gene dynamics.

Acknowledgements. Research supported by the program P1-0044 (Slovenia) and national project OI141035 (Serbia), and international projects BI-RS/08-09-047 and MRTN-CT-2004-005728. We used the computer system of the Department of theoretical physics, Jožef Stefan Institute, Ljubljana, and on the AEGIS e-Infrastructure, supported in part by EU FP6 and FP7 projects CX-CMCS, EGEE-III and SEE-GRID-SCI, at the Scientific Computing Lab, Institute of Physics, Belgrade.

References

1. Boccaletti, S., Latora, V., Moreno, Y., Chavez, M., Hwang, D.-U.: Complex networks: Structure and dynamics. Physics Reports 424, 175–308 (2006)
2. Tavazoie, S., Hughes, J.D., Campbell, M.J., Cho, R.J., Church, G.M.: Nature Genet. 22, 281 (1999); Li, F., Long, T., Lu, Y., Ouyang, Q., Tang, C.: Proc. Natl. Acad. Sci. 101, 4781 (2004); Balcan, D., Erzan, A.: Eur. Phys. J. B 38, 253 (2004)
3. Mantegna, R.N.: Hierarchical structure in financial markets. European Physical Journal B 11, 193–197 (1999)
4. Takayasu, M., Takayasu, H., Sato, T.: Critical behavior and 1/f noise in information traffic. Physica A 233, 824–834 (1996)
5. Tadić, B., Rodgers, G.J., Thurner, S.: Transport on Complex Networks: Flow, Jamming & Optimization. Int. J. Bifurcation and Chaos 17(7), 2363–2385 (2007)
6. Abel, M., Ahnert, K., Kurths, J., Mandelj, S.: Additive nonparametric reconstruction of dynamical systems from time series. Phys. Rev. E 71(1), 015203 (2005)
7. Lockhart, D.J., Winzeler, E.A.: Genomics, gene expression and DNA arrays. Nature 405, 827–836 (2000)
8. Gardner, T.S., Faith, J.J.: Reverse-engineering transcription control networks. Physics of Life Reviews 2, 65–88 (2005)
9. Stokić, D., Hanel, R., Thurner, S.: A fast and efficient gene-networks reconstruction methodfrom multiple overexpression experiments (2008) ArXiv e-prints
10. Tadić, B., Mitrović, M.: Jamming and correlation patterns in traffic of information on sparse modular networks (2009) ArXiv e-prints
11. Fortunato, S., Barthelemy, M.: Resolution limit in community detection. PNAS 104, 36–41 (2007)
12. Danon, L., Diaz-Guilera, A., Arenas, A.: Effect of size heterogeneity on community identification in complex networks. J. Stat. Mechanics: Theory & Experiment, P11010 (2006)
13. Cho, R.J., Campbell, et al.: A Genome-Wide Transcriptional Analysis of the Mitotic Cell Cycle. Molecular Cell 2, 65–73 (1998), `http://arep.med.harvard.edu/cgi-bin/ExpressDByeast`
14. Caretta-Cartozo, C., De Los Rios, P., Piazza, F., Lio, P.: Bottleneck Genes and Community Structure in the Cell-Cycle Network of S.pombe. PLoS Comput. Biol. 3(6), e103 (2007)
15. Živkovic, J., Tadić, B., Wick, N., Thurner, S.: Statistical Indicators of Collective Behavior and Functional Clusters in Gene Expression Network of Yeast. European Physical Journal B 50, 255 (2006)
16. Mitrović, M., Tadić, B.: Spectral and dynamical properties in classes of sparse networks with mesoscopic inhomogeneity (2008) arxiv.org:0809.4850
17. Madi, A., Friedman, Y., Roth, D., Regev, T., Bransburg-Zabary, S., Jacob, E.B.: Genome holography: Deciphering function-form motifs from gene expression data. PLoS ONE 3(7), e2708 (2008)
18. Eisler, Z., Kertesz, J.: Random walks on complex networks with inhomogeneous impact. Physical Review E 71(5), 057104 (2005)
19. `http://mips.gsf.de`
20. Tsallis, C.: J. Stat. Phys. 52, 479 (1988)
21. Zipf, G.K.: Psycho-Biology of Languages. Houghton-Mifflin, Cambridge (1935) (MIT Press, 1965)
22. Furusawa, C., Kaneko, K.: Phys. Rev. Lett. 90, 088102 (2003)
23. Farkas, I.J., Derényi, I., Barabási, A.-L., Vicsek, T.: Spectra of real-world graphs: Beyond the semicircle law. Phys. Rev. E 64(2), 026704 (2001)

Detecting and Characterizing the Modular Structure of the Yeast Transcription Network

Alexandre S. Cristino, Roberto F.S. Andrade, and Luciano da F. Costa

Abstract. Systems biology and complex networks research will turn biology into a more precise and synthetic discipline. To date, complex network concepts have been used to study all diversity of networks, such as social organization to molecular interaction. In this study we are particularly interested in addressing some aspects of the structural and functional organization of biological networks. The construction of a comprehensive regulatory map of molecular systems will contribute to a better understanding of the 'design principles' of the genetic regulatory networks. We proposed a reliable strategy in blending bioinformatics and complex network research for characterization of modular structure in sparse biological networks with low average node degree, as for the yeast transcription network (YTN). We find that the YTN is highly modular and those modules have specific functions. In addition, communities or modules sharing structural properties are also sharing some functional traits, which is a remarkable finding. Our approach could be used for helping biologists to address specific biological questions by designing hypothesis-driven experiments.

Alexandre S. Cristino
Institute of Physics of São Carlos, University of São Paulo, São Carlos, SP, Brazil, PO Box 369, 13560-970, Phone +55 16 3373 9879, Fax +55 16 3373 9879
e-mail: `alexsc@rge.fmrp.usp.br`

Roberto F.S. Andrade
Institute of Physics, Federal University of Bahia, Campus de Ondina, 40210-340, Salvador, Brazil, Phone +55 71 3283 6657, Fax +55 71 3283 6607 and National Institute of Science and Technology for Complex Systems, Brazil
e-mail: `randrade@ufba.br`

Luciano da F. Costa
Institute of Physics of São Carlos, University of São Paulo, São Carlos, SP, Brazil, PO Box 369, 13560-970, phone +55 16 3373 9858, fax +55 16 3371 3616 and National Institute of Science and Technology for Complex Systems, Brazil
e-mail: `luciano@if.sc.usp.br`

R. Menezes et al. (Eds.): Complex Networks: CompleNet 2009, SCI 207, pp. 35–46.
springerlink.com © Springer-Verlag Berlin Heidelberg 2009

1 Introduction

Though the reductionist approach to biology and medicine has led to several important advances, further progresses with respect to the remaining challenges require integration of representation, characterization and modeling of the studied systems along a wide range of spatial and time scales. Such an approach, intrinsically related to systems biology, is poised to ultimately turning biology into a more precise and synthetic discipline, paving the way to extensive preventive and regenerative medicine [1], drug discovery [20] and treatment optimization [24].

A particularly appealing and effective approach to addressing the complexity of interactions inherent to the biological systems is provided by the new area of complex networks [34, 30, 8, 13, 12]. Basically, it is an extension of graph theory [10], focusing on the modeling, representation, characterization, analysis and simulation of complex systems by considering many elements and their interconnections. Complex networks concepts and methods have been used to study disease [17], transcription networks [5, 6, 4], protein-protein networks [22, 36, 16, 39], metabolic networks [23] and anatomy [40].

Three basic concepts are critical for the proper understanding of complex biological systems: emergence, robustness and modularity. Complex systems display emergent properties that are not explained by their individual parts alone [1]. Robustness is an inherent property of the biological systems strongly favored by evolution to maintain their phenotypic traits stable even through different environments, stochastic events and genetic variation [1]. Finally, modularity is a critical level of biological organization that contributes to both robustness of the entire system by confining failure to individual parts or modules, and to evolution, by rearranging modules [19]. Modules can be understood as 'building blocks' of the biological systems that interact in order to shape the phenotypic traits of the cells and organisms [19].

The great complexity of organisms arises more as a consequence of elaborated regulation of gene expression than from differences in genetic content in terms of the number of genes [9, 26]. Of particular interest are the eukaryotic transcription networks that regulate the transcription of many genes and will be ultimately translated in functional proteins determining the molecular changes in the cellular environment [41]. Transcription regulation maps have been constructed for *Saccharomyces cerevisiae* [25, 18, 27]. The interactions between transcription factors (TFs) and target genes (TGs) can be represented as a directed graph. The two types of nodes (TF and TG) are connected when regulatory interaction occurs between regulators and targets [5, 2]. These networks are organized in substructures such as motifs and modules. Modularity in these networks arises from groups of genes assigned to particular biological functions, but there is no consensus on the precise groups of genes and interactions scaffolding modular structure [21, 7, 5]. The systematic investigation of transcriptional networks therefore constitutes a subject of great interest for researchers in both systems biology and complex networks.

Communities or modules are observed in many real-world networks and can be defined as set of nodes which are more intensely connected one another than with the remainder of the network [40]. Various methods for detecting communities have been developed, such as spectral [35], hierarchical [37] and modularity-base [33, 31] approaches. After identifying the communities in a given biological network, it becomes interesting to quantify and investigate their specific topological features. Such a study can provide insights not only about the structural homogeneity between the communities, but also yield insights about possible relationships between specific topological properties and respective functionality. One of the first comprehensive studies of the individual properties of communities was reported with respect to bone canals system [40, 14]. After obtaining a 3D reconstruction of the bone canal network for a whole phalange bone (chicken), that structure was transformed into a graph, which was found to contain well-defined communities. Each of those communities was isolated from the rest of the graph and had several topological properties (e.g. degree, clustering coefficient) quantified. Considerable differences were found between the several bone communities, indicating a gradient of structural organization along the longitudinal axis of the bone, which is potentially related to mechanical function and/or development. Although that approach is suitable for applications such as for bone networks, it is not viable when the networks have low average degree such as in the case with many biological systems. Indeed, many interesting biological networks present modular structure, but low average degree imposes a problem in using the previous developed method in the sense that rather sparse networks are obtained after separating each community from the remainder of the original network.

The current work addresses two important problems in systems biology, namely (i) the mapping of gene regulatory networks using bioinformatics analysis and databases, and (ii) the study of relationships among communities found in those networks, as quantified by their respective topological features. Once the communities have been identified, it is possible to take into account ontological information of the gene function, as defined by the Gene Ontology Consortium [3], in order to identify the main functionalities of each community. Because the topological modules (i.e. communities) are potentially related to function, it is possible that each module have specific main biological roles. At the same time, the relationship between the structural properties of each community can lead to additional insights about how the system of modules is interconnected and can be potentially related to specific biological processes. For instance, would two or more communities with similar topological measurements present similar functional traits? These possibilities are explored here with respect to the most complete transcriptional regulatory network of the yeast to date [27], yielding several interesting results such as the identification of communities with clear main biological functionalities as well as the identification of specific structural-functional relationships between the communities. Also, we proposed a possibility to overcome the limitation of making measurements of each community while still attached to the original structure. This approach is different from the methods previously described regarding bone canals system [40, 14] and is a new contribution for measuring structural properties of communities while connected to the major network rather than separately.

We start by presenting the data and methods, and proceed by reporting and discussing the results obtained for the yeast transcriptional regulatory network. Many biological systems have been represented as complex networks but a few have been characterized at different hierarchical levels as in the case of communities in *S. cerevisae* transcription network.

2 Material and Methods

The transcriptional regulatory network of *Saccharomyces cevevisiae*, which is studied in this manuscript, was created by mapping regulatory sites conserved in the genome of other three species of yeast (*S. paradoxus*, *S. mikatae* and *S. bayannus*), as described elsewhere [27]. In that work, the authors proposed an improved map of transcriptional control in yeast by reanalysing chromatin immunoprecipitation data for 203 transcription factors with more conservative criteria. The network was constructed by linking a transcription factor (TF) to a target that can be a TF or other types of proteins. The interaction between nodes represents a transcriptional regulation of a TF over another protein that can be a TF. The original yeast transcription network contains 117 transcription factors [27]. This network is composed of seven connected components, but only the largest one was considered in this study since the other six include no more than four TFs. The major connected component has 2069 nodes, including 113 transcription factors and 1956 regulated genes. Commonly, a bipartite network would be used to represent this type of biological network. However, we represent it as a more general regulatory network in which both transcription factors and the regulated genes are considered as genes. As it is possible to have interactions between transcription factors, we resourced to a network representation involving just one type of nodes, i.e. genes, therefore naturally accommodating the interconnections between transcription factors.

The resulting structure is henceforth called the yeast transcription network (YTN). All gene identification numbers were standardized to the official locus name as defined in *Saccharomyces* Genome Database (SGD - http://www.yeastgenome.org).

In order to find the communities in the YTN, we used an algorithm that is based on the maximization of the modularity, a measurement of the degree of separation among the communities [11], as implemented in the igraph package (http://cneurocvs.rmki.kfki.hu/igraph/). The topology of each community was characterized and analysed in terms of the three following measurements: average degree, clustering coefficient and average shortest distance between the nodes in each community [13].

Python programming language was used in order to integrate programs and databases containing the network, cross-references for gene name identification, gene functional annotation, as well as network properties and measurements. Cytoscape software [38] was used for edition and visualization of the network topology and properties.

A new strategy was used in this work, which is suitable for studying topological properties of communities in networks with low average degree, as in the case of the

YTN (Figure 1A). The method consists in calculating and analysing the topological properties of each community without separating it from the original network, differently from the previously adopted approach [40, 14]. Functional annotation for biological process of yeast genes was obtained from the Gene Ontology database [3] (GO - http://www.geneontology.org).

BiNGO software [28] was used in order to perform this functional analysis for the communities found in the YTN. The functional analysis is a hypergeometric test that quantifies in terms of P-value how much a functional category is enriched in a given set of genes compared to the whole reference set, in our case the entire network.

3 Results

Figure 1A shows the graph representation of the YTN, in which nodes correspond to genes and the edges between nodes indicate that there is a regulatory control between those genes. The YTN contains 2069 nodes including 113 transcription factors and 1956 regulated genes, as well as 4100 interconnecting edges. This network has low average degree ($k = 3.95$) and clustering coefficient ($cc = 0.0124$).

We looked for modular organization in the YTN to be subsequently related to biological function. A modularity value above 0.3 is generally considered as a strong evidence of modular organization in the network [29, 32]. The YTN yielded modularity equals to 0.62, so that YTN can be considered a highly modular network, with 15 well-defined communities (Figure 1B). Except for community 11, all other communities turned out to belong to specific functional roles (Table 1). Also, these communities were compared by their structural properties (Table 1). Those

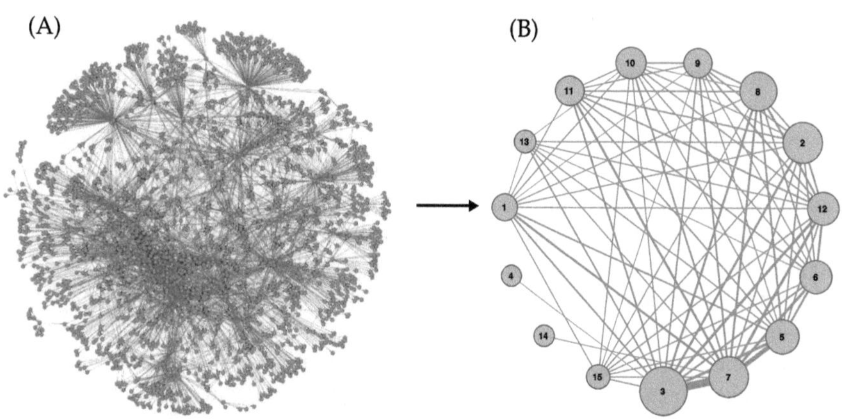

Fig. 1 Modular organization of the yeast transcription network (YTN). (A) *S. cerevisiae* transcription network as proposed by MacIsaac *et al.* 2006 [27]. (B) The 15 communities in the YTN were found by using Clauset's et al. algorithm [11]. The node size is proportional to the number of genes in the respective community, while the edge width is proportional to the number of connections among communities

Table 1 Structural and functional properties of the communities identified in the YTN. TF, number of transcription factors; f_{TF}, frequency of transcription factors in the community; k, average degree; cc, clustering coefficient; ℓ, average shortest distance between nodes

Community	Nodes	Edges	TF	f_{TF}	k	cc	ℓ	Biological process	P-value
1	79	111	4	0.051	3.96	0.155	2.85	response to stress	3.57E-04
2	249	259	6	0.024	2.62	0.024	2.80	catabolic process	4.98E-05
3	346	762	18	0.052	5.47	0.182	3.10	response to pheromone DNA replication	1.68E-07 1.19E-06
4	9	8	1	0.111	1.89	0.000	1.58	vitamin metabolic process	2.86E-06
5	170	266	13	0.076	4.35	0.055	3.36	lipid metabolic process ion transport	3.86E-05 1.33E-03
6	157	201	14	0.089	3.55	0.086	3.22	cellular biosynthetic process	4.54E-06
7	251	492	19	0.076	5.63	0.120	3.03	hydrogen transport	1.77E-02
8	221	248	9	0.041	2.93	0.003	3.01	organelle organization and biogenesis	2.40E-02
9	108	110	4	0.037	2.93	0.003	2.44	phosphate transport	3.18E-02
10	134	197	8	0.060	3.50	0.001	2.74	biosynthetic process	4.19E-11
11	109	117	6	0.055	3.18	0.017	3.37	–	–
12	151	206	6	0.040	3.93	0.047	2.75	glycolysis organic acid metabolic process	5.83E-05 2.80E-03
13	27	28	2	0.074	2.85	0.004	2.64	vitamin metabolic process cofactor metabolic process	2.64E-03 2.64E-03
14	12	13	2	0.167	2.25	0.225	1.65	regulation of transcription by carbon catabolites response to nutrient	2.71E-06 3.74E-06
15	46	46	1	0.022	2.63	0.015	1.91	spore wall assembly	6.09E-06

communities are interconnected by 1036 edges and, except for modules 4 and 14, all modules are densely interconnected (Figure 1B). The number of connections between each community is different and community 3 and 7 are those which are most connected in the YTN (Figure 1B). The average and standard deviation of the number of nodes and edges are equal to 138±98 and 204±200, respectively.

Figure 2A presents the scatter plot obtained for 15 communities while taking into account the three adopted measurements (i.e. average degree, clustering coefficient and average shortest distance between nodes inside each community). These communities can be grouped with respect to their relative distances in the multidimensional space (Figure 2A). Note that two or more communities, which are close one another in this space, will exhibit similar topological measurements, and therefore similar connectivity [13]. Such a proximity between communities allows us to explore the interaction between those modules in order to ultimately build the whole functional genetic system for transcriptional control in yeast cells. Also, we searched for over-representation of functional GO categories in the 7 identified clusters of communities. Interestingly, most biological processes over-represented

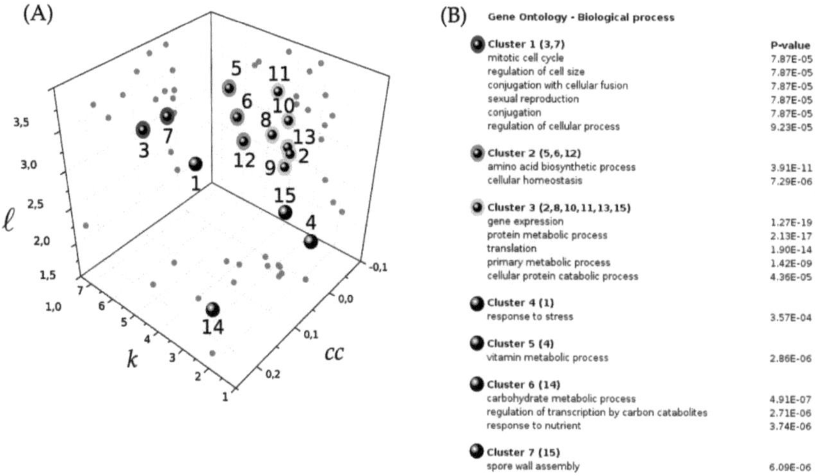

Fig. 2 Analysis of the structural properties of the communities identified in the YTN (A). Seven clusters of communities were identified while taking into account their proximity in the 3D scatterplot in (A). These clusters seem to share functional roles in yeast (B). ℓ, average shortest distance between nodes; k, average node degree; cc, clustering coeficient

in these clusters of genes are of great importance for regulation of critical process in a biological system such as the yeast cell (Figure 2B).

Figure 3 shows the clusters of communities identified in terms of their relative proximity in the measurement space and their assigned functional role in cellular biology. The YTN has many functional modules mainly related to cell cycle, reproduction, response to changes in the environment, gene expression, degradation process, cellular homeostasis, cellular structure and metabolic process (Figure 3).

4 Discussion

The YTN was found to be composed by 15 communities related to many different biological processes (Table 1). The proportion of transcription factors in each community ranges from 2% to 16%, and most of the communities have no more than 10% of transcription factor, which means that a few number of factors are capable to regulate many different biological processes. Most functional modules in the YTN are densely interconnected, which means an intense cross-talking between biological processes. Two communities, 4 and 14, are connected only once to the communities 3 and 7, respectively. Interestingly, communities 3 and 7 are the biggest and most important modules since they are involved in the cell cycle and reproduction. Also, they are highly connected to each other and are connected to most communities in the YTN (Figure 3).

Clusters of communities were identified in the YTN based on their structural properties and, surprisingly, communities with similar connectivity patterns tended

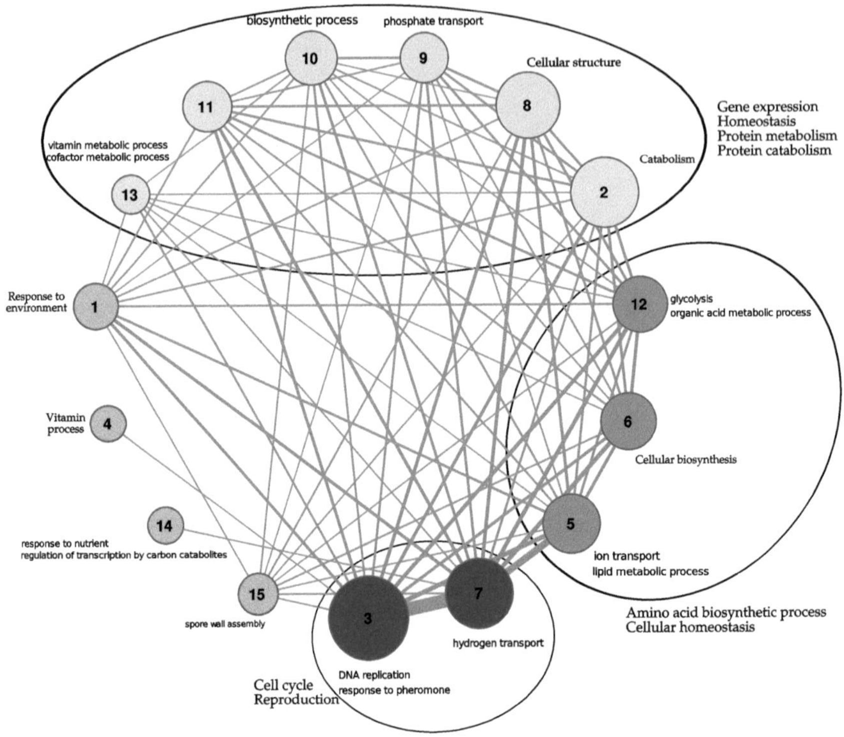

Fig. 3 The yeast transcription network is organised as structural modules of transcription factors and target genes that are related to various specific functions

to share functional traits (Table 1, Figure 2). These results are particularly interesting because communities with similar structures should not necessarily have to perform similar functions. This is in accordance to the idea of 'building blocks' of the biological systems that ultimately shape the phenotypic traits of an organism [19]. Although the method used here for finding communities in the networks [11] has limitations in the resolution of the modularity function [15], we were still able to find robust and biologically relevant results concerning the structural properties of communities and their funcional roles.

Those communities have interesting functional and structural traits that are worth discussing. For instance, cluster 1, which is involved in cell cycle and reproduction, is composed by community 3 that works on DNA replication and response to pheromone, while community 7 is involved in hydrogen transport (red nodes in Figure 3). Cluster 2 controls the amino acid biosynthesis and cellular homeostasis, while communities 5 is related to ion transport and lipid metabolism, community 6 plays role in the cellular biosynthesis, and community 12 is related to glycosysis and organic acid metabolism (green nodes in Figure 3). Also, cluster 3 must be crucial for cellular homeostasis, gene expression, protein metabolism and degradation which has sub-modules for catabolism (community 2), cellular structure (community 8), phosphate transport (community 9), biosynthetic process (community 10) and vitamin and cofactor metabolism (community 13) (yellow nodes, Figure 3). The other four cluster are composed by one community each and also have important functions for the *S. cerevisae*, such as response to environment, vitamin process, response to the environment and spore wall assembly (blue nodes, Figure 3).

Further studies on the structural patterns in those community could provide useful insights for understanding the principles of these biological circuits, helping in the design of biosynthetic systems. As many genes are still lacking functional characterization, our approach could be useful in assigning functional roles to those genes based on the structural and functional traits of their respective communities.

5 Conclusions

The modular organization of biological networks had been previously identified in yeast transcription networks based on genome-wide location and gene expression data [21, 7] by using clustering algorithm rather than complex networks measurements. However, an objective measurement of the modularity of the interactions was still lacking. We used a reliable algorithm [11] to identify communities in the YTN, which was constructed based on the most complete *S. cerevisae* transcriptional regulatory network [27]. The YTN provides a representation of genes and their transcriptional regulatory relationship. Given that this network is sparse, we developed a reliable strategy for studying communities in this type of networks by detecting and measuring multiple features of modules without separating them from the original network.

We found that the YTN is highly modular (modularity = 0.62) and 15 communities were identified for this network. Furthermore, analysis of the topological properties of the obtained communities indicated that some modules share one or more structural properties with other modules. Communities and cluster of communities were assigned to functional roles based on the over-representation of specific biological function in those sub-networks. Our results show that modules structurally similar also tend to share functional properties. The seven clusters of communities identified in our approach represent a potential core of the maintenance of the cellular processes. These clusters of modules have more general functions than the isolated communities. The different levels of functional specificity in the communities and clusters of communities can be an evidence of a hierarchical organization in biological systems.

In fact, the network structure and topology of the YTN can be related to functional roles in the cell system. We provided a reliable method for studying modules in a sparse network and could be used to assign potential functions to novel genes still poorly characterized by functional assays. In addition, our results can help to better understand how the genetic systems can be modularized in such a way that the diversity of biological patterns could emerge and be shaped by evolution. It would be interesting to analyse in detail each of those communities in order to check for evolutionary conservation in the modular organization of *E. coli* transcription network.

Acknowledgements. Alexandre S. Cristino is grateful to FAPESP (proc. 06/61232-2) for financial support. Roberto F. S. Andrade was partially supported by CNPq (306052/2007-5). Luciano da F. Costa acknowledges FAPESP (05/00587-5) and CNPq (301303/2006-1 and 573583/2008-0) for financial support. We also gratefully acknowledge Francisco A. Rodrigues, Lucas Antiqueira, Matheus P. Viana, Bruno Traveolo and Paulino R. Villas-Boas for their comments about this manuscript.

References

1. Aderem, A.: Systems Biology: Its Practice and Challenges. Cell 121(4), 511–513 (2005)
2. Alon, U.: An Introduction to Systems Biology: Design Principles of Biological Circuits. Chapman and Hall/CRC (2007)
3. Ashburner, M., Ball, C.A., Blake, J.A., Botstein, D., Butler, H., Cherry, J.M., Davis, A.P., Dolinski, K., Dwight, S.S., Eppig, J.T., Harris, M.A., Hill, D.P., Issel-Tarver, L., Kasarskis, A., Lewis, S., Matese, J.C., Richardson, J.E., Ringwald, M., Rubin, G.M., Sherlock, G.: Gene ontology: tool for the unification of biology. the gene ontology consortium. Nat. Genet. 25, 25–29 (2000)
4. Babu, M.M.: Computational approaches to study transcriptional regulation. Biochem. Soc. Trans. 36(Pt 4), 758–765 (2008)
5. Babu, M.M., Luscombe, N.M., Aravind, L., Gerstein, M., Teichmann, S.A.: Structure and evolution of transcriptional regulatory networks. Curr. Opin. Struct. Biol. 14(3), 283–291 (2004)
6. Balaji, S., Babu, M.M., Aravind, L.: Interplay between network structures, regulatory modes and sensing mechanisms of transcription factors in the transcriptional regulatory network of E. coli. J. Mol. Biol. 372(4), 1108–1122 (2007)
7. Bar-Joseph, Z., Gerber, G.K., Lee, T.I., Rinaldi, N.J., Yoo, J.Y., Robert, F., Gordon, D.B., Fraenkel, E., Jaakkola, T.S., Young, R.A., Gifford, D.K.: Computational discovery of gene modules and regulatory networks. Nat. Biotechnol. 21(11), 1337–1342 (2003)
8. Barabasi, A.L., Oltvai, Z.N.: Network biology: understanding the cell's functional organization. Nat. Rev. Genet. 5(2), 101–113 (2004)
9. Carroll, S.B.: Endless forms: the evolution of gene regulation and morphological diversity. Cell 101, 577–580 (2000)
10. Chartrand, G., Lesniak, L.: Graphs & digraphs. Wadsworth Publ. Co. Belmont (1986)
11. Clauset, A., Newman, M.E.J., Moore, C.: Finding community structure in very large networks. Phys. Rev. E 70(6), 066111–066117 (2004)
12. da Fontoura Costa, L., Rodrigues, F.A., Cristino, A.S.: Complex networks: The key to systems biology. Genetics and Molecular Biology 31(3), 591–601 (2008)
13. da Fontoura Costa, L., Rodrigues, F.A., Travieso, G., Villas-Boas, P.R.: Characterization of complex networks: A survey of measurements. Advances in Physics 56(1), 167–242 (2007)
14. da Fontoura Costa, L., Viana, M.P., Beletti, M.E.: The complex channel networks of bone structure. Appl. Phys. Lett. 88, 33903 (2006)
15. Fortunato, S., Barthelemy, M.: Resolution limit in community detection. Proc. Natl. Acad. Sci. U S A 104(1), 36–41 (2007)

16. Giot, L., Bader, J.S., Brouwer, C., Chaudhuri, A., Kuang, B., Li, Y., Hao, Y.L., Ooi, C.E., Godwin, B., Vitols, E., Vijayadamodar, G., Pochart, P., Machineni, H., Welsh, M., Kong, Y., Zerhusen, B., Malcolm, R., Varrone, Z., Collis, A., Minto, M., Burgess, S., McDaniel, L., Stimpson, E., Spriggs, F., Williams, J., Neurath, K., Ioime, N., Agee, M., Voss, E., Furtak, K., Renzulli, R., Aanensen, N., Carrolla, S., Bickelhaupt, E., Lazovatsky, Y., DaSilva, A., Zhong, J., Stanyon, C.A., Finley Jr., R.L., White, K.P., Braverman, M., Jarvie, T., Gold, S., Leach, M., Knight, J., Shimkets, R.A., McKenna, M.P., Chant, J., Rothberg, J.M.: A protein interaction map of Drosophila melanogaster. Science 302(5651), 1727–1736 (2003)
17. Goh, K.I., Cusick, M.E., Valle, D., Childs, B., Vidal, M., Barabasi, A.L.: The human disease network. Proc. Natl. Acad. Sci. U S A 104(21), 8685–8690 (2007)
18. Harbison, C.T., Gordon, D.B., Lee, T.I., Rinaldi, N.J., Macisaac, K.D., Danford, T.W., Hannett, N.M., Tagne, J.B., Reynolds, D.B., Yoo, J., Jennings, E.G., Zeitlinger, J., Pokholok, D.K., Kellis, M., Rolfe, P.A., Takusagawa, K.T., Lander, E.S., Gifford, D.K., Fraenkel, E., Young, R.A.: Transcriptional regulatory code of a eukaryotic genome. Nature 431(7004), 99–104 (2004)
19. Hartwell, L.H., Hopfield, J.J., Leibler, S., Murray, A.W.: From molecular to modular cell biology. Nature 402(suppl. 6761), C47–C52 (1999)
20. Hopkins, A.L.: Network pharmacology: the next paradigm in drug discovery. Nat. Chem. Biol. 4(11), 682–690 (2008)
21. Ihmels, J., Friedlander, G., Bergmann, S., Sarig, O., Ziv, Y., Barkai, N.: Revealing modular organization in the yeast transcriptional network. Nat. Genet. 31(4), 370–377 (2002)
22. Jeong, H., Mason, S.P., Barabási, A.-L., Oltvai, Z.N.: Lethality and centrality in protein networks. Nature 411(6833), 41–42 (2001)
23. Jeong, H., Tombor, B., Albert, R., Oltvai, Z.N., Barabasi, A.L.: The large-scale organization of metabolic networks. Nature 407(6804), 651–654 (2000)
24. Kitano, H.: Computational systems biology. Nature 420(6912), 206–210 (2002)
25. Lee, T.I., Rinaldi, N.J., Robert, F., Odom, D.T., Bar-Joseph, Z., Gerber, G.K., Hannett, N.M., Harbison, C.T., Thompson, C.M., Simon, I., Zeitlinger, J., Jennings, E.G., Murray, H.L., Gordon, D.B., Ren, B., Wyrick, J.J., Tagne, J.B., Volkert, T.L., Fraenkel, E., Gifford, D.K., Young, R.A.: Transcriptional regulatory networks in saccharomyces cerevisiae. Science 298(5594), 799–804 (2002)
26. Levine, M., Tjian, R.: Transcription regulation and animal diversity. Nature 424, 147–151 (2003)
27. MacIsaac, K.D., Wang, T., Gordon, D.B., Gifford, D.K., Stormo, G.D., Fraenkel, E.: An improved map of conserved regulatory sites for Saccharomyces cerevisiae. BMC Bioinformatics 7, 113 (2006)
28. Maere, S., Heymans, K., Kuiper, M.: BiNGO: a Cytoscape plugin to assess overrepresentation of gene ontology categories in biological networks. Bioinformatics 21(16), 3448–3449 (2005)
29. Newman, M.E.: Modularity and community structure in networks. Proc. Natl. Acad. Sci. U S A 103(23), 8577–8582 (2006)
30. Newman, M.E.J.: The Structure and Function of Complex Networks. SIAM Review 45(2), 167–256 (2003)
31. Newman, M.E.J.: Fast algorithm for detecting community structure in networks. Phys. Rev. E 69, 066133 (2004)
32. Newman, M.E.J.: Detecting community structure in networks. Phys. Rev. E 38(2), 321–330 (2006)
33. Newman, M.E.J., Girvan, M.: Finding and evaluating community structure in networks. Phys. Rev. E 69, 026113 (2004)

34. Oltvai, Z.N., Barabasi, A.L.: Systems biology. Life's complexity pyramid. Science 298(5594), 763–764 (2002)
35. Pothen, A., Simon, H., Liou, K.-P.: Partitioning sparse matrices with eigenvectors of graphs. SIAM J. Matrix Anal. Appl. 11, 430 (1990)
36. Rain, J.C., Selig, L., De Reuse, H., Battaglia, V., Reverdy, C., Simon, S., Lenzen, G., Petel, F., Wojcik, J., Schaechter, V., et al.: The protein-protein interaction map of Helicobacter pylori. Nature 409(6817), 211–215 (2001)
37. Scott, J.: Social network analysis: a handbook. Sage, London (2000)
38. Shannon, P., Markiel, A., Ozier, O., Baliga, N.S., Wang, J.T., Ramage, D., Amin, N., Schwikowski, B., Ideker, T.: Cytoscape: a software environment for integrated models of biomolecular interaction networks. Genome. Res. 13(11), 2498–2504 (2003)
39. Simonis, N., Rual, J.F., Carvunis, A.R., Tasan, M., Lemmens, I., Hirozane-Kishikawa, T., Hao, T., Sahalie, J.M., Venkatesan, K., Gebreab, F., Cevik, S., Klitgord, N., Fan, C., Braun, P., Li, N., Ayivi-Guedehoussou, N., Dann, E., Bertin, N., Szeto, D., Dricot, A., Yildirim, M.A., Lin, C., de Smet, A.S., Kao, H.L., Simon, C., Smolyar, A., Ahn, J.S., Tewari, M., Boxem, M., Milstein, S., Yu, H., Dreze, M., Vandenhaute, J., Gunsalus, K.C., Cusick, M.E., Hill, D.E., Tavernier, J., Roth, F.P., Vidal, M.: Empirically controlled mapping of the Caenorhabditis elegans protein-protein interactome network. Nat. Methods 6(1), 47–54 (2009)
40. Viana, M.P., Tanck, E., Beletti, M.E., da Fontoura Costa, L.: Modularity and robustness of bone networks. Mol. Biosyst (in Press, 2009), doi:10.1039/b814188f
41. Wray, G.A., Hahn, M.W., Abouheif, E., Balhoff, J.P., Pizer, M., Rockman, M.V., Romano, L.A.: The evolution of transcriptional regulation in eukaryotes. Mol. Biol. Evol. 20, 1377–1419 (2003)

Finding Overlapping Communities Using Disjoint Community Detection Algorithms

Steve Gregory

Abstract. Many algorithms have been designed to discover community structure in networks. Most of these detect disjoint communities, while a few can find communities that overlap. We propose a new, two-phase, method of detecting overlapping communities. In the first phase, a network is transformed to a new one by splitting vertices, using the idea of *split betweenness*; in the second phase, the transformed network is processed by a disjoint community detection algorithm. This approach has the potential to convert any disjoint community detection algorithm into an overlapping community detection algorithm. Our experiments, using several "disjoint" algorithms, demonstrate that the method works, producing solutions, and execution times, that are often better than those produced by specialized "overlapping" algorithms.

1 Introduction and Motivation

Networks are a natural representation for various kinds of complex system, in society, biology, and other fields. One of the most interesting properties of many types of network is their *community structure*: the existence of groups, or *communities*, of vertices that are more densely connected to each other than to vertices in other communities. Communities often represent related groups of individuals in the real world. The automatic discovery of network communities is very useful because, for example, it can help throw light on the structure of networks which are far too large for humans to make sense of manually, even with the help of visualization techniques.

There is currently no generally accepted definition of *community*, and no standard algorithm exists for discovering communities. Numerous algorithms, using a variety of methods, have been developed; these vary in their effectiveness and speed for different types of network. Many algorithms were described in the survey papers of [8, 19], but recently many new algorithms have appeared, including

Steve Gregory
Department of Computer Science, University of Bristol, Bristol BS8 1UB, England
e-mail: steve@cs.bris.ac.uk

R. Menezes et al. (Eds.): Complex Networks: CompleNet 2009, SCI 207, pp. 47–61.
springerlink.com © Springer-Verlag Berlin Heidelberg 2009

some very fast ones with the potential to work on very large networks. Most community detection (CD) algorithms assume that networks are unipartite and have undirected, unweighted edges; we make the same assumptions in this paper.

An important difference between algorithms is their view of the relation between the communities in a network. The vast majority of algorithms, including [6, 11, 15, 20, 25, 27, 29], assume that vertices are members of a *flat* set of *disjoint* communities. This makes sense for many networks: for example, most employees work for a single employer, most papers are published in a single conference, etc. A few algorithms, including [2, 12, 13, 24, 30], allow communities to overlap, with each individual possibly appearing in more than one community. This is more realistic in some cases: for example, many researchers belong to more than one research community. Yet other algorithms [3, 5, 17] aim to detect a *hierarchy* of communities: for example, a number of research communities each divided into several research groups.

The dichotomy between "disjoint" and "overlapping" CD algorithms is unfortunate because it limits the application of each algorithm. If a network has overlapping communities, a "disjoint" algorithm cannot find them; conversely, if communities are known to be disjoint, a "disjoint" algorithm will generally perform better than an "overlapping" algorithm. For the best results for a given network, it is important to use the right kind of algorithm. (The question of how to choose the right kind of algorithm is outside the scope of the present paper.)

In this paper we present a method to allow *any* "disjoint" CD algorithm to be used instead for finding overlapping communities. This means that a user wishing to find overlapping communities need no longer be forced to use one of the small number of "overlapping" algorithms that exist, but can also choose from the many "disjoint" algorithms. Moreover, improved "disjoint" algorithms resulting from future research can potentially also be applied to the problem of detecting overlapping communities.

Our method is implemented by transforming a network into another network that can be fed into a "disjoint" CD algorithm, and then transforming the resulting disjoint communities into (potentially overlapping) communities of the original network. The transformation is based on the *split betweenness* principle introduced in the CONGA CD algorithm [12, 13].

The next section provides a brief overview of the CONGA algorithm, which inspired this work. In Section 3 we present our transformation algorithm, named Peacock, explain its design, and compare it with CONGA. Section 4 describes the results of experiments to detect overlapping communities in both synthetic and real-world networks. The experiments use a combination of Peacock with four existing "disjoint" algorithms, as well as two existing "overlapping" algorithms. Conclusions appear in Section 5.

2 The CONGA Algorithm

CONGA (Cluster-Overlap Newman Girvan Algorithm) [12] is a CD algorithm based on Girvan and Newman's [11, 22] "GN" algorithm but extended to detect overlapping communities. CONGA adds to the GN algorithm the ability to split vertices between communities, based on the new concept of split betweenness.

CONGA comprises a sequence of steps, each of which removes an edge from the network *or* splits a vertex into two vertices:

1. Calculate edge betweenness of edges and split betweenness of vertices.
2. Remove edge with maximum edge betweenness or split vertex with maximum split betweenness, if greater.
3. Recalculate edge betweenness and split betweenness.
4. Repeat from step 2 until no edges remain.

The *edge betweenness* [10, 11] of an edge *e* is the number of shortest paths, between all pairs of vertices, that pass along *e*. The *split betweenness* [12] of a vertex *v* is the number of shortest paths that would pass between the two parts of *v* if it were split. There are many ways to split a vertex into two; the *best split* is the one that maximizes the split betweenness. Ref. [12] gives an approximate, efficient algorithm for calculating split betweenness at the same time as edge betweenness.

In CONGA, a network is initially treated as a single community, assuming it is connected. After one or more iterations, step 2 causes the network to split into two components (communities). Communities are repeatedly split into two until only singleton communities remain. By representing the binary splits as a dendrogram, the network can be partitioned into any desired number of communities.

The algorithm has a worst-case time complexity of $O(n^3)$ for a sparse network. In practice, the speed depends on the number of vertices that are split (which increases the network size) and on how easily the network breaks into separate components. This is because, in step 3, betweenness need be calculated only for the component containing the removed edge or split vertex, or for both components if step 2 caused the component to split.

Ref. [13] presents an optimized version of CONGA, named CONGO (CONGA Optimized), which employs a *local* form of betweenness. In CONGO, edge betweenness and split betweenness are calculated by counting the number of *short* paths: those that are no longer than *h* (a parameter). This optimization reduces the time complexity to $O(n \log n)$ for a sparse network. For simplicity, we refer to both CONGA and CONGO by the name CONGA in the remainder of this paper.

3 The Peacock Algorithm

The Peacock algorithm is used in the context shown in Fig. 1. The system comprises the following phases:

1. The network is transformed to a new, larger, network. Each step of the transformation *splits* a vertex into two vertices and one edge. Assuming the original network was connected, the transformed one will also be connected. The names of the vertices involved in each splitting step are stored for later use; for example, if vertex *v* splits into $\{v, v'\}$, v' is recorded as a copy of *v* in the vertex

names file. Additionally, all vertices in the transformed file are renamed to integers, for compatibility with some CD algorithms that impose this restriction.
2. The transformed network is input to a CD algorithm, which produces a *clustering*: a set of disjoint sets of vertices.
3. The disjoint clustering is converted to a (possibly overlapping) clustering by replacing the vertex names by those used in the original network. For example, if Peacock split v into $\{v, v'\}$ and these occur in two different sets in the disjoint clustering, the final clustering includes v in both sets, which therefore overlap.

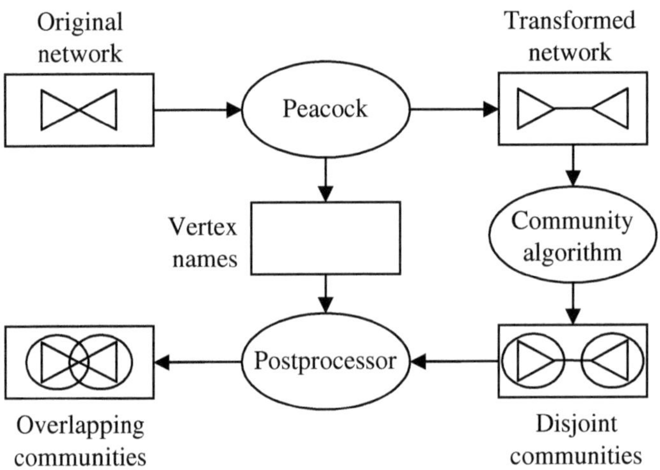

Fig. 1 Architecture of the Peacock system

The Peacock algorithm itself, which transforms the network, works as follows:

1. Calculate the split betweenness of all vertices.
2. Choose the vertex with the maximum split betweenness. Split it into two, according to its best split.
3. Recalculate the split betweenness of vertices, where this might have changed.
4. Repeat from step 2 until the maximum split betweenness is sufficiently small.
5. For each split vertex, place a new edge between the two resulting vertices.

In step 4, termination depends on the ratio between the maximum split betweenness (of all vertices) and the maximum edge betweenness (of all edges). The loop terminates when this ratio becomes less than s, a parameter of the algorithm.

In step 5, edges are placed in the order in which vertices split. For example, if v splits twice, creating v' and v'', edges $\{v,v'\}$ and $\{v,v''\}$ are placed; if v' then splits twice, creating v''' and v'''', $\{v',v'''\}$ and $\{v',v''''\}$ are placed, as shown in Fig. 3(a).

Fig. 2 shows an example of Peacock's transformation of a simple network. Fig. 2(a) is the original network. The maximum split betweenness is 40, for vertex a,

(a)

(b)

(c)

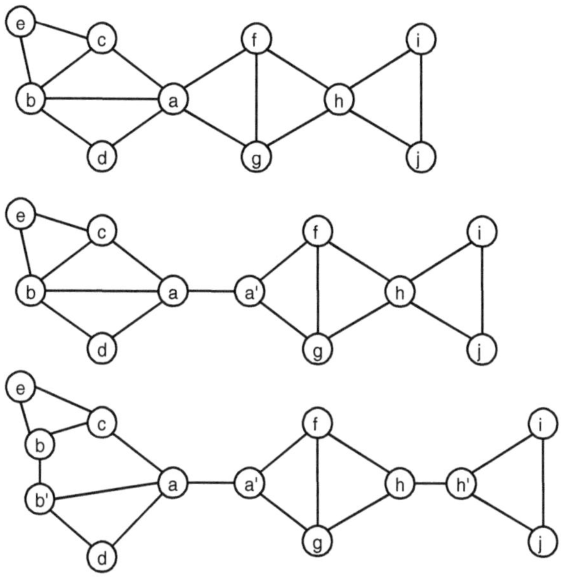

Fig. 2 Example of Peacock network transformation: (a) original network; (b) after first splitting step; (c) after all three splitting steps

while the maximum edge betweenness is 25, for edges {a,f} and {a,g}. Provided the *s* parameter is less than 1.6 (=40/25), a will be split. Its best split is ({b,c,d},{f,g}), so the network will be transformed to that shown in Fig. 2(b). If *s* is small enough (less than 0.8) there are two more splitting steps that can be done, splitting h and then b, resulting finally in the network of Fig. 2(c).

Provided the *s* parameter is small enough, its exact value is not critical to the result of the community detection. For example, if we ask a CD algorithm to divide the network of Fig. 2(b) into two disjoint communities, the result will usually be {{a,b,c,d,e}, {a',f,g,h,i,j}}. If we feed the network of Fig. 2(c) into the same algorithm, it will find larger communities, {{a,b,b',c,d,e}, {a',f,g,h,h',i,j}}, but both of these solutions are postprocessed to the same pair of overlapping communities: {{a,b,c,d,e}, {a,f,g,h,i,j}}. We return to the choice of *s* below in this section.

The Peacock algorithm is quite similar to CONGA. One difference is that CONGA does not bridge the gaps formed when a vertex splits. Another difference is that CONGA interleaves the vertex splitting steps (as described above) with edge removal steps (as in the GN algorithm [11]). In CONGA, both vertex splitting and edge removal steps act to break down a network into separate components which represent communities. Peacock is not intended to detect communities, and so it keeps the network connected.

Design Alternatives

We choose to recalculate betweenness in each iteration, instead of simply splitting the vertices that initially have a split betweenness greater than a certain value. This is for the same reason as the GN and CONGA algorithms recalculate betweenness in each iteration: the network structure changes, breaking into separate components, and the values of betweenness rapidly become out of date. Besides, sometimes a vertex needs to be split more than once; its betweenness after its first split cannot be calculated at the beginning.

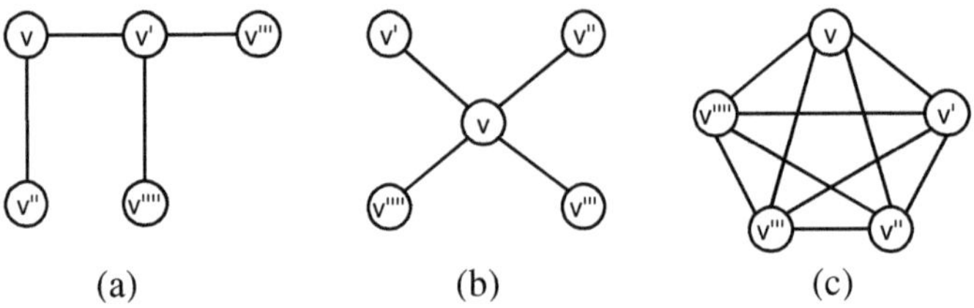

Fig. 3 Alternative ways to connect split vertices: (a) method used in Peacock; (b) connecting vertex to each of its copies; (c) connecting vertices in a clique

Another key design decision is whether, and how, to bridge the gaps formed when vertices split. The method used is to place an edge across each gap as it is created. For example, if v splits twice, creating v' and v'', and v' splits twice, creating v''' and v'''', the edges placed are shown in Fig. 3(a). This method was chosen because the edge betweenness of the new edge approximately equals the split betweenness of the split vertex. (It is not identical because the network contains extra vertices and longer paths following the split.) This helps make the community structure apparent in the transformed network fed to the CD algorithm. The following alternatives were also considered:

1. Do not add any edges. The problem with this method is that the network is likely to break into disconnected components during the transformation process, which affects the communities that can be found by the CD algorithm.
2. Place an edge only when necessary to prevent the network splitting into two components. This avoids the above problem, but the network still breaks into almost-separate components connected by few edges, so the results are poor.
3. Place an edge between the original vertex (v) and each of the copies of it (v', v'', v''', v''''), as in Fig. 3(b). This method gives worse results than the chosen one.
4. Join the original vertex and its copies in a clique, as in Fig. 3(c). This method sometimes works well but is very sensitive to the value of parameter s. This is because a clique, especially a large one, is treated as one community by most CD algorithms, so there is no advantage in transforming a vertex to a clique.

Another issue is the value of parameter s. In most networks, the maximum split betweenness is slightly greater than the maximum edge betweenness, so setting s to a value greater than about 1 or 2 leaves the network unchanged. A smaller value of s causes more vertices to be split and the network to increase in size. We have experimented with s ranging from 0.005 to 0.5, and found remarkably little difference in the solution quality. Some CD algorithms favour a larger value while some prefer a smaller value, but the difference is small. As regards execution time, a large value of s is preferable, so that the CD algorithm will have a smaller network to process. We settled on a value of s=0.1 for all experiments in the next section.

The final design decision is the value of the parameter h. For CONGA, reducing h usually reduces both solution quality and execution time. Using Peacock, combined with the GN algorithm, we varied h for each phase: reducing h for Peacock had a much smaller effect on solution quality than reducing h for the GN algorithm. The same is true of the other CD algorithms. This suggests that local betweenness is a more acceptable optimization for "splitting" than for community detection. We therefore used h=2 for all experiments in the next section.

4 Experiments

To evaluate Peacock, we combined it with several disjoint CD algorithms. These were chosen because they are modern algorithms with the potential to handle large networks, *and* implementations of them, by their authors, were readily available:

1. CNM. Clauset, Newman, and Moore's "fast modularity" algorithm of [6, 31].
2. WT. The algorithm of Wakita and Tsurumi [27, 32] (rev. 159): an optimization of the CNM algorithm.
3. BGLL. The "fast unfolding" algorithm of Blondel, Guillaume, Lambiotte, and Lefebvre [3, 33] (February 2008 version): another modularity-maximizing algorithm, claimed to be faster than CNM or WT.
4. PL. The "Walktrap" algorithm of Pons and Latapy [25, 34] (v0.2), which works by generating random walks which tend to get trapped in communities.

We compare the results with results from two existing overlapping CD algorithms, whose code is also available:

1. CFinder. The "clique percolation" algorithm of Palla, Derényi, Farkas, and Vicsek [1, 24, 35] (v1.21).
2. CONGA. Gregory's CONGA algorithm [12, 13, 36] (v1.59) with h=2.

For CNM, WT, PL, and CONGA, the user can choose the desired number of communities, although for WT there is a minimum number of communities that can be found. In contrast, BGLL and CFinder find a small number of solutions, each with a fixed number of communities.

Experiments with Synthetic Networks

A good way to evaluate a CD algorithm is by generating artificial networks based on a known community structure and comparing the known communities with those found by the algorithm. The comparison can be done in various ways, including the F-measure and Mutual Information measure [9]. The Adjusted Rand index [16], a variant of the Rand index [26] that excludes the effects of chance, is often considered the most accurate. However, it is not ideal for solutions containing overlapping clusters because it does not consider the number of clusters containing each pair of vertices. We therefore use the Omega index [7]: an extension of the Adjusted Rand index for solutions with overlapping clusters.

We randomly generated a set of networks containing n vertices divided into c equally-sized communities, each containing nr/c vertices. Vertices are randomly and evenly distributed between communities so that each vertex is a member of r communities on average. r is a measure of overlap: r=1 means that communities are disjoint and r=c means that each community contains all vertices. The network is constructed by placing edges between pairs of vertices randomly, with probability ip_{in} if there are i ($\geq$1) communities to which both vertices belong, and p_{out} otherwise. All networks used in the experiments are connected. Results shown are the average of 100 runs.

In these experiments we evaluate Peacock (h=2) combined with CNM, WT, and PL, and compared these with CONGA (h=2) and CFinder. For most of the algorithms we ask for c communities, where c is the known number of communities in the network. This is impossible with CFinder, whose only parameter is k (cluster density), so we show the results from CFinder for *all* values of k.

Fig. 4 shows results for 256 vertices in 32 communities. The overlap is 2, so each community contains 16 vertices. As p_{out} increases, the community structure becomes less evident and the solution quality decreases, more sharply for CONGA than for CFinder. Peacock+PL behaves similarly to CONGA, but Peacock+CNM is much better – comparable with CFinder – while Peacock+WT is slightly worse.

Fig. 5 shows the effect of increasing the density of intracommunity edges, which should increase the solution quality. All combined algorithms perform better than CONGA for low p_{in}, with Peacock+WT slightly worse than the others.

In Fig. 6 we fix p_{in} and p_{out} and vary the overlap, r. CONGA's performance declines as r increases above 2. Peacock+PL behaves slightly better while Peacock+CNM is better than CONGA or CFinder. Again, Peacock+WT performs slightly less well than Peacock+CNM.

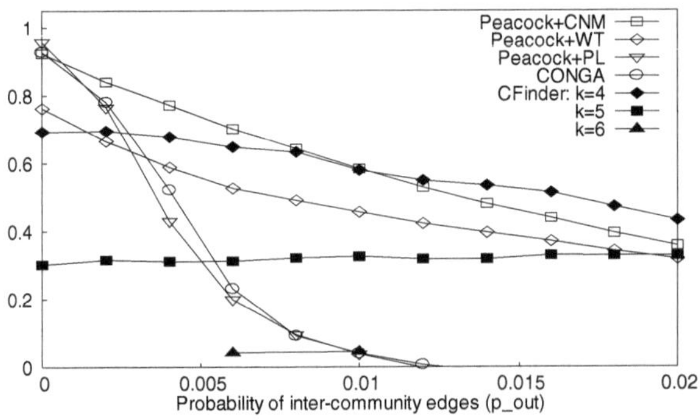

Fig. 4 Omega index for random networks with n=256, c=32, r=2, p_{in}=0.5, various p_{out}

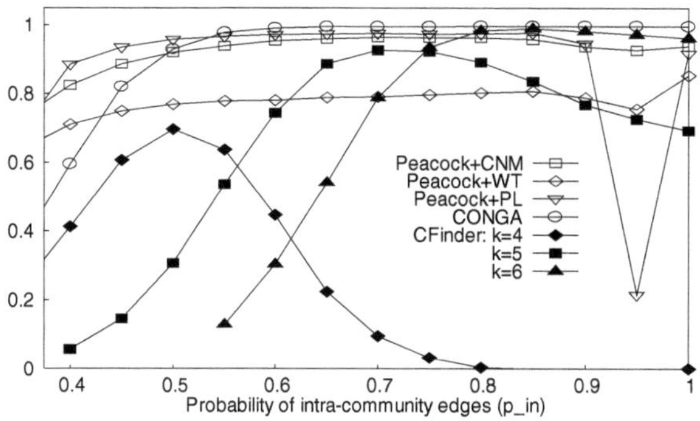

Fig. 5 Omega index for random networks with n=256, c=32, r=2, p_{out}=0, various p_{in}

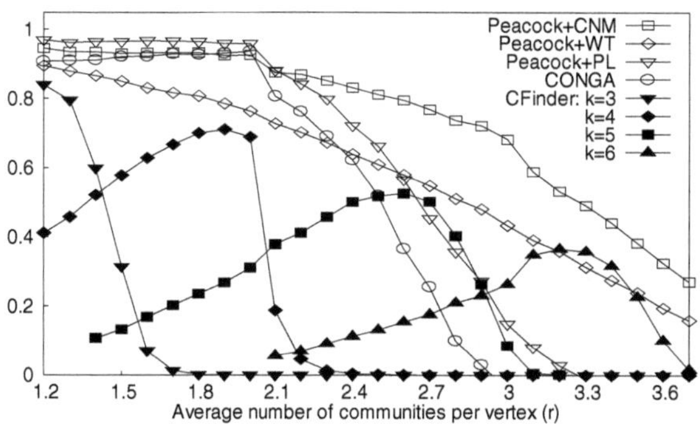

Fig. 6 Omega index for random networks with n=256, c=32, p_{in}=0.5, p_{out}=0, various r

Fig. 7 shows the effect of varying the network size while keeping the community size constant. Peacock+CNM and Peacock+PL both perform better than CONGA, with Peacock+WT slightly worse than the others.

In Fig. 8 the network size is fixed but the number (and therefore size) of the communities varies. Peacock+CNM performs better than CONGA; the other combined algorithms perform slightly worse, but still better than CFinder.

Fig. 9 shows how the total execution time, to detect the specified number of communities, varies with network size. All these networks contain overlapping communities of a small fixed size with overlap 1.2. All programs were run under Linux on an AMD Opteron 250 CPU at 2.4GHz. For each of the combined "Peacock+X" algorithms, the execution time plotted comprises the time for the network-transformation phase (using Peacock) *plus* the time for the CD phase (using algorithm X). The time for the Peacock phase is similar to CONGA's execution time: it increases almost linearly with size, at least for the experiment shown here. The time for the CD phase depends on which algorithm is used *and* how sensitive that algorithm is to network size, since it has to process the transformed network, which is larger than the original one whose size is shown on the horizontal axis.

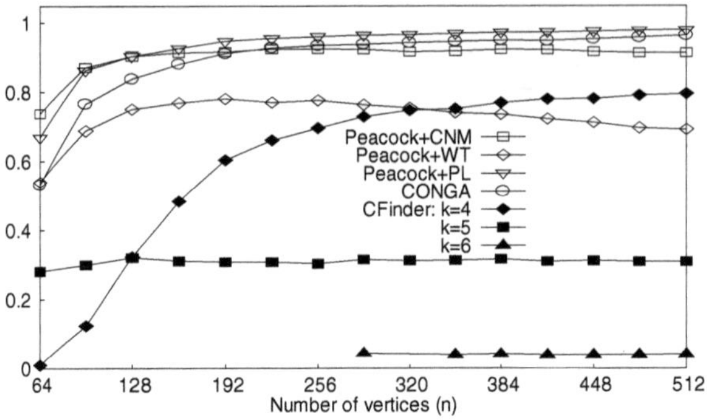

Fig. 7 Omega index for random networks with c=n/8, r=2, p_{in}=0.5, p_{out}=0, various n

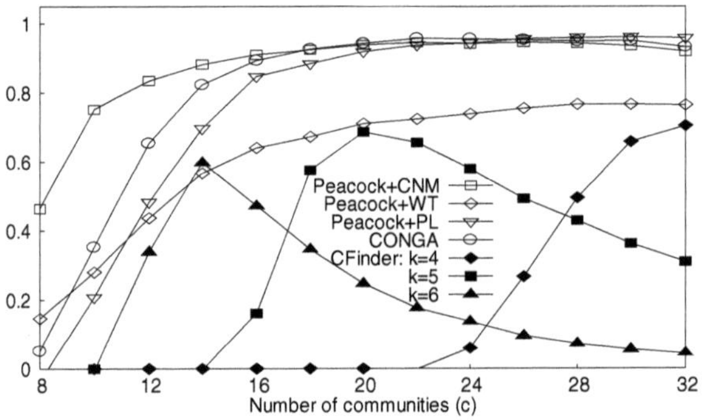

Fig. 8 Omega index for random networks with n=256, r=2, p_{in}=0.5, p_{out}=0, various c

This shows that, for large networks, execution time is greatest for CFinder and least for CONGA, while Peacock+WT is the fastest of the combined algorithms.

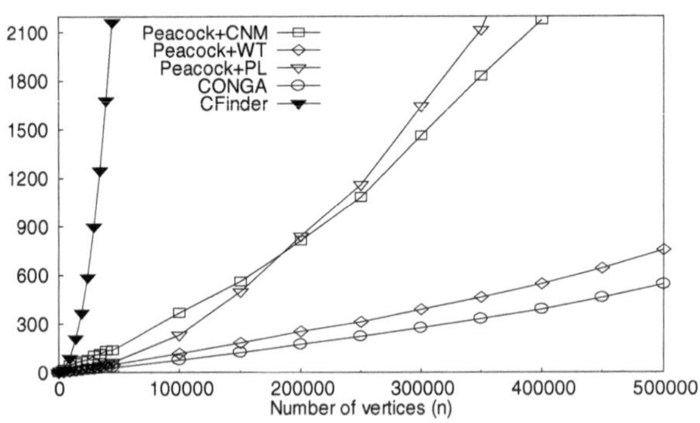

Fig. 9 Execution time (seconds) for random networks c=n/8, r=1.2, p_{in}=0.5, p_{out}=0, various n

Experiments with Real-World Networks

We have run the CD algorithms on several real-world networks, listed in Table 1. The table shows the source of each network, its size, and the times for the various algorithms to generate solutions, on an AMD Opteron 250 at 2.4GHz. The total execution time is the sum of the execution time of the Peacock network-transformation phase (column 7) and the time of the CD phase (last four columns). Again, the time for the Peacock phase is similar to CONGA's execution time.

When evaluating a CD algorithm on real-world networks, there is usually no known "correct" solution. Solution quality must be assessed in a different way: for example, by modularity [21, 22], which measures the relative number of intra-community and intercommunity edges. A high modularity indicates that there are more intracommunity edges than would be expected by chance.

The original modularity measure, Q, is defined only for disjoint communities, but Nicosia *et al.* [23] proposed a new modularity measure, Q_{ov}, which *is* defined also for overlapping communities. Q_{ov} is defined so that Q_{ov}=0 when all vertices belong to one community or all belong to singleton communities, while higher values of Q_{ov} indicate stronger community structure. Each vertex may belong to each community with any *belonging coefficient*. For each vertex, the belonging coefficients for all communities sum to 1.

Table 1 Results on real-world networks

Name	Ref	Vertices	Edges	Execution time (s)						
				CONGA	CFinder	Peacock: combined algorithm				
						Peacock	Phase 2			
							CNM	WT	PL	BGLL
netscience	20	379	914	1.3	0.25	0.35	0.6	0.9	0.03	0.06
cond-mat-2003	18	27519	116181	1127	1134	1088	82.8	10.8	24.1	3.71
blogs	28	3982	6803	6.5	3.05	5.23	6.74	2.15	0.43	0.32
blogs2	28	30557	82301	294	415	289	86.3	9.8	33.9	∞
PGP	4	10680	24316	83	34745	89.8	18.6	4.62	2.66	0.90
email	14	1133	5451	30.2	4.00	32.6	4.00	2.61	0.45	0.24
word_association	24	7205	31784	175	96.5	176	33.2	4.47	6.49	1.78
protein-protein	24	2445	6265	8.6	2.75	7.56	4.84	2.00	0.41	∞

We use modularity (Q_{ov}) here to evaluate solutions on real-world networks. The belonging coefficient of each vertex is set to $1/c$, where c is the number of communities it belongs to; i.e., vertices belong equally to all communities they are in.

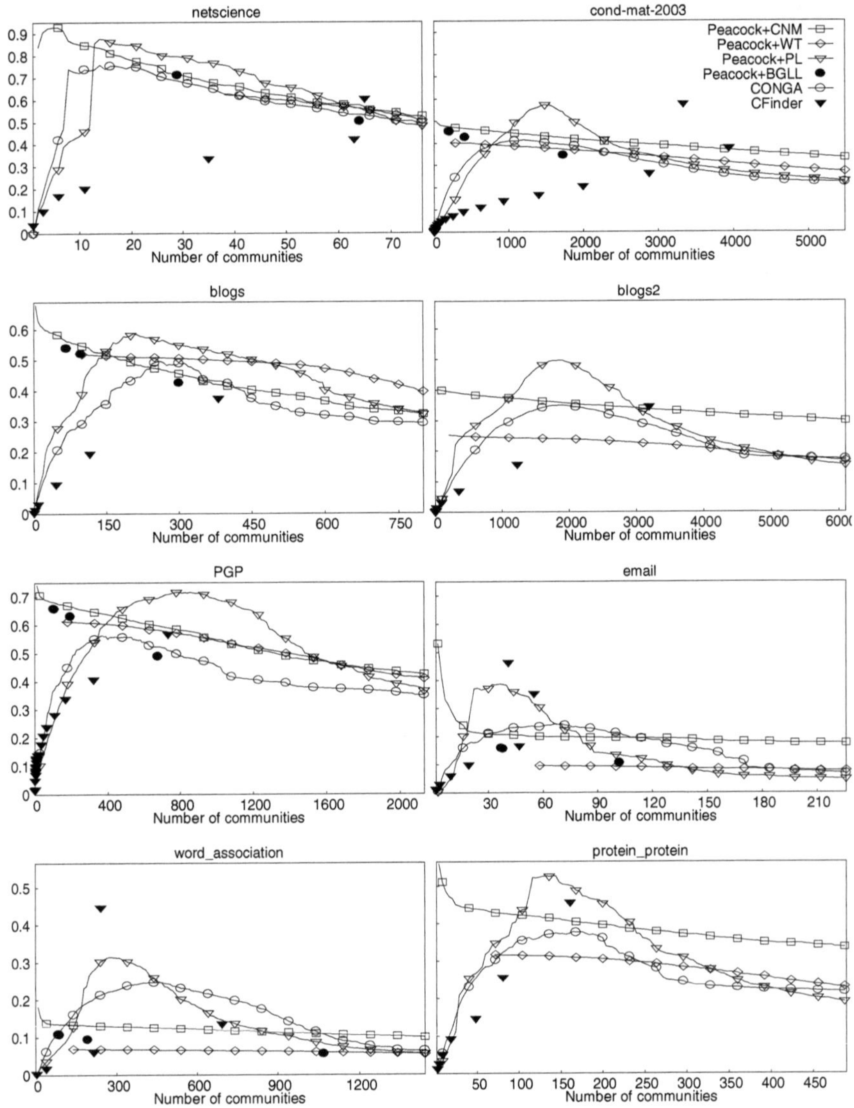

Fig. 10 Modularity of real-world networks. The *y*-axis shows the Q_{ov} modularity

Fig. 10 shows the modularity of the eight networks listed in Table 1. "netscience" and "cond-mat-2003" are collaboration networks of coauthorships, of different sizes. Peacock+PL finds the solutions with the highest modularity, over a certain range; otherwise, the best results are obtained by Peacock+CNM. Both give a higher modularity than CONGA. CFinder finds several solutions, one of which has a slightly higher modularity than the other algorithms.

"blogs" and "blogs2" are networks of communication relationships between owners of blogs on the MSN (Windows Live™) Spaces website. "blogs2" is much larger than "blogs" and has a higher average degree. "PGP" and "email" are other social networks representing PGP key signing and email, respectively. For all four networks, the story is the same as for "netscience" and "cond-mat-2003": Peacock+PL gives the best results over a certain range, Peacock+CNM gives consistently good results, and both perform better than CONGA. For the "blogs" network, Peacock+WT also does well.

The last two, "word_association" and "protein_protein", are non-social networks, from psychology and biology, respectively, both from [24]. For the first of these, Peacock+PL finds a higher-modularity solution than CONGA, but not quite as good as CFinder's best solution. For the second, Peacock+PL finds even better solutions than CFinder's best, while Peacock+CNM also performs well.

5 Conclusions

We have proposed a novel, two-phase, approach to detecting overlapping communities in networks. In principle, this is attractive because it separates the "overlapping" and "community detection" issues, allowing the best algorithm to be selected for each phase. For the first phase, we have presented the Peacock algorithm, based on the "split betweenness" principle. While the CONGA algorithm uses the betweenness principle for both overlapping and community detection, Peacock uses it only for the former. Interestingly, the local form of betweenness [13] works better in Peacock than it does in CONGA, suggesting that our approach should yield good execution speed.

The results reported in Section 4 seem to confirm that our approach is viable. In terms of solution quality, the two-phase algorithm works well, especially with the CNM or PL algorithm as the second phase. In most cases, these combined algorithms outperform the two specialized "overlapping" CD algorithms.

Concerning execution time, for the small networks shown in Table 1, the time for the combined algorithms is dominated by the Peacock phase, whose execution time is similar to CONGA's, while CFinder's execution time is usually better but sometimes worse. For larger networks, as Fig. 7 shows, CFinder's execution time increases rapidly with network size, while the time for the 2-phase algorithm becomes dominated by the second (community detection) phase. The time for the Peacock phase, as for CONGA, increases almost linearly with size. For large networks, most of the total execution time is occupied by the CD phase, unless this is done by the (fast) WT algorithm. The time for the CD phase varies according to the algorithm used, but will always be longer than for detecting disjoint communities, because the algorithm needs to process the larger, transformed, network.

Future work includes evaluating the Peacock algorithm in conjunction with even more disjoint CD algorithms, including those that have yet to be designed. The implementation of the Peacock algorithm, including its postprocessor, is available at http://www.cs.bris.ac.uk/~steve/networks/ .

References

1. Adamcsek, B., Palla, G., Farkas, I., Derényi, I., Vicsek, T.: CFinder: Locating Cliques and Overlapping Modules in Biological Networks. Bioinformatics 22, 1021–1023 (2006)
2. Baumes, J., Goldberg, M., Magdon-Ismail, M.: Efficient Identification of Overlapping Communities. In: Kantor, P., Muresan, G., Roberts, F., Zeng, D.D., Wang, F.-Y., Chen, H., Merkle, R.C. (eds.) ISI 2005. LNCS, vol. 3495, pp. 27–36. Springer, Heidelberg (2005)
3. Blondel, V.D., Guillaume, J.-L., Lambiotte, R., Lefebvre, E.: Fast Unfolding of Communities in Large Networks. J. Stat. Mech., P10008 (2008)
4. Boguña, M., Pastor-Satorras, R., Diaz-Guilera, A., Arenas, A.: Models of Social Networks Based on Social Distance Attachment. Phys. Rev. E 70, 056122 (2004)
5. Clauset, A., Moore, C., Newman, M.E.J.: Hierarchical Structure and the Prediction of Missing Links in Networks. Nature 453, 98–101 (2008)
6. Clauset, A., Newman, M.E.J., Moore, C.: Finding Community Structure in Very Large Networks. Phys. Rev. E 70, 066111 (2004)
7. Collins, L.M., Dent, C.W.: Omega: A General Formulation of the Rand Index of Cluster Recovery Suitable for Non-disjoint Solutions. Multivar. Behav. Res. 23, 231–242 (1988)
8. Danon, L., Diaz-Guilera, A., Duch, J., Arenas, A.: Comparing Community Structure Identification. J. Stat. Mech., P09008 (2005)
9. Fred, A.L.N., Jain, A.K.: Robust Data Clustering. In: IEEE Computer Society Conference on Computer Vision and Pattern Recognition, pp. 128–133. IEEE Press, New York (2003)
10. Freeman, L.C.: A Set of Measures of Centrality Based on Betweenness. Sociometry 40, 35–41 (1977)
11. Girvan, M., Newman, M.E.J.: Community Structure in Social and Biological Networks. P. Natl. Acad. Sci. USA 99, 7821–7826 (2002)
12. Gregory, S.: An Algorithm to Find Overlapping Community Structure in Networks. In: Kok, J.N., Koronacki, J., Lopez de Mantaras, R., Matwin, S., Mladenič, D., Skowron, A. (eds.) PKDD 2007. LNCS (LNAI), vol. 4702, pp. 91–102. Springer, Heidelberg (2007)
13. Gregory, S.: A Fast Algorithm to Find Overlapping Communities in Networks. In: Daelemans, W., Goethals, B., Morik, K. (eds.) PKDD 2008. LNCS (LNAI), vol. 5211, pp. 408–423. Springer, Heidelberg (2008)
14. Guimera, R., Danon, L., Diaz-Guilera, A., Giralt, F., Arenas, A.: Self-similar Community Structure in a Network of Human Interactions. Phys. Rev. E 68, 065103(R) (2003)
15. Hofman, J.M., Wiggins, C.H.: Bayesian Approach to Network Modularity. Phys. Rev. Lett. 100, 258701 (2008)
16. Hubert, L., Arabie, P.: Comparing partitions. J. Classif. 2, 193–218 (1985)
17. Lancichinetti, A., Fortunato, S., Kertesz, J.: Detecting the Overlapping and Hierarchical Community Structure of Complex Networks (2008) Eprint arXiv:0802.1218v1 at arxiv.org
18. Newman, M.E.J.: The Structure of Scientific Collaboration Networks. P. Natl. Acad. Sci. USA 98, 404–409 (2001)
19. Newman, M.E.J.: Detecting Community Structure in Networks. Eur. Phys. J. B 38, 321–330 (2004)

20. Newman, M.E.J.: Finding Community Structure in Networks Using the Eigenvectors of Matrices. Phys. Rev. E 74, 036104 (2006)
21. Newman, M.E.J.: Modularity and Community Structure in Networks. P. Natl. Acad. Sci. USA 103, 8577–8582 (2006)
22. Newman, M.E.J., Girvan, M.: Finding and Evaluating Community Structure in Networks. Phys. Rev. E 69, 026113 (2004)
23. Nicosia, V., Mangioni, G., Carchiolo, V., Malgeri, M.: Extending Modularity Definition for Directed Graphs with Overlapping Communities (2008) Eprint arXiv:0801.1647v3 at arxiv.org
24. Palla, G., Derényi, I., Farkas, I., Vicsek, T.: Uncovering the Overlapping Community Structure of Complex Networks in Nature and Society. Nature 435, 814–818 (2005)
25. Pons, P., Latapy, M.: Computing Communities in Large Networks Using Random Walks. J. Graph Algorithms and Applications 10(2), 191–218 (2006)
26. Rand, W.M.: Objective Criteria for the Evaluation of Clustering Methods. J. Am. Stat. Assoc. 66, 846–850 (1971)
27. Wakita, K., Tsurumi, T.: Finding Community Structure in a Mega-scale Social Networking Service. In: IADIS International Conference on WWW/Internet 2007, pp. 153–162 (2007)
28. Xie, N.: Social Network Analysis of Blogs. MSc Dissertation. University of Bristol (2006)
29. Xu, X., Yuruk, N., Feng, Z., Schweiger, T.A.: SCAN: a Structural Clustering Algorithm for Networks. In: 13th International Conference on Knowledge Discovery and Data Mining, KDD 2007, pp. 824–833. ACM, New York (2007)
30. Zhang, S., Wang, R., Zhang, X.: Identification of Overlapping Community Structure in Complex Networks Using Fuzzy C-means Clustering. Physica A 374(1), 483–490 (2007)
31. Clauset, A.: `http://cs.unm.edu/~aaron/research/fastmodularity.htm`
32. Wakita, K.: `http://www.is.titech.ac.jp/~wakita/en/software/community-analysis-software`
33. Guillaume, J-L.: `http://findcommunities.googlepages.com`
34. Pons, P.: `http://psl.pons.free.fr/index.php?item=prog&lang=en`
35. CFinder: `http://www.cfinder.org`
36. Gregory, S.: `http://www.cs.bris.ac.uk/~steve/networks`

Discovering Community Structure on Large Networks Using a Grid Computing Environment

Dario Fiumicello, Alessandro Longheu, and Giuseppe Mangioni

Abstract. Natural arising and evolution of community structures in natural and social networks has been explained as a result of topological relationships among nodes in the same network, and many studies in this field have revealed that it is possible to derive information about the community decomposition of a network just by examining its structure. The most used metric for this kind of analysis is the so-called "modularity" [12][11], which expresses the quality of a candidate community decomposition of a network. Despite its popularity, modularity is hard to be optimized [2] and algorithms for communities discovering based on modularity optimization are practically unfeasible for large networks. On the other hand, methods for community uncovering based on locally evaluated metric are very fast [7]. In this paper we propose the use of a parallel implementation of the local metric based method for community discovering proposed in [9] and the use of the overlapping modularity function [13] to evaluate the best partition. All measures reported in this paper are obtained running our implementation within a grid computing environment.

1 Introduction

The concept of community is very familiar in real life. Almost all people somehow belong to one of the communities that naturally arise when a set of people interact in

Dario Fiumicello
Dipartimento di Ingegneria Informatica e delle Telecomunicazioni,
University of Catania, Italy
e-mail: dario.fiumicello@tiscali.it

Alessandro Longheu
Dipartimento di Ingegneria Informatica e delle Telecomunicazioni,
University of Catania, Italy
e-mail: alongheu@diit.unict.it

Giuseppe Mangioni
Dipartimento di Ingegneria Informatica e delle Telecomunicazioni,
University of Catania, Italy
e-mail: gmangioni@diit.unict.it

R. Menezes et al. (Eds.): Complex Networks: CompleNet 2009, SCI 207, pp. 63–71.
springerlink.com © Springer-Verlag Berlin Heidelberg 2009

a social context and especially when they share common interests (hobbies, politics, etc.). Communities have been extensively studied by sociologists and psychologists for their importance in social dynamics, and in the last few years, by people involved in other scientific fields (physics, mathematics, computer science, ...)[1][10].

Despite our familiarity with these concepts, it is not so easy to define in a formal way what a community is. One of the most widely used definition is the one given by Newman and Girvan in [12]: *a community is a set of nodes more densely linked to each other than to the rest of nodes or, equivalently, a graph has a community structure if the number of the links into any subgraph is higher than the number of links between those subgraphs.*

Starting from this definition, several different algorithms for community detection have been proposed (see [4][7] for an overview). Many of them are based on the optimization of a function that gives a measure of the quality of the community structure. The most famous is the modularity function defined by Newman ([12][11]). Even if an important resolution limit of the modularity measure has been pointed out by Fortunato and Barthelemy in [6], the modularity seems to be an useful quality measure for a network community partition. This measure however is not appropriate when the network cannot be divided into a set of sharp communities, which is a very common situation especially in social networks where a person usually belongs to several different communities at the same time. For this reason, the definition of modularity has been further extended in order to capture overlapping structures ([13]). Despite its popularity, modularity is hard to be optimized [2] and algorithms for communities discovering based on modularity optimization are practically unfeasible for large networks. On the other hand, methods for community uncovering based on locally evaluated metric are very fast and in literature there exist several examples of their applicability to large real networks ([7]). In particular, in [9] a method (*L*ancichinetti-*F*ortunato-*K*ertesz *m*ethod - $LFKm$ for the sake of brevity) based on a local fitness function for overlapping and hierarchical community structure detecting is presented. This method is able to discover not only the overlapping communities, but also the whole network hierarchical structure. Indeed, the output of this algorithm is a set of partitions (defined as a set of clusters such that each node is assigned to at least one cluster), obtained tuning a resolution parameter. In order to discover the best (i.e. stable) partition, it is necessary to compare partitions obtained at different resolution scales. Following the authors' approach, each partition is evaluated using the average value of the fitness ($f_{\mathscr{P}}$) of its communities and the best partition is decided plotting $f_{\mathscr{P}}$ and looking at the length of the plateaus. However, we have experimented that for large networks with no sharp community division, plateaus are hard to be detected and, in many cases, they are not present at all. Despite this, the approach seems very fast, so our idea is to combine such a local metric based method with a global measure as, for example, the modularity. In this way, the best partition corresponds to the maximum of the modularity evaluated for each partition. Since nodes could belong to several communities (i.e. overlapped nodes), we capture the best division using the modularity for overlapped communities Q_{ov}.

Moreover, to further improve the performance of *LFKm*, we implemented a faster version of this method, suitable to be executed in a grid computing environment in a parallel fashion. The results obtained are very satisfying and, as shown in detail in this paper, very few current algorithms are able to analyze networks of such large size in such a short time.

The outline of the paper is as follows: in section 2 a short introduction of *LFKm* is given; in section 3 our implementation of *LFKm* is presented and, finally, in section 4 results of the application of the presented approach to well-known networks are shown.

2 *LFKm*: A Brief Introduction

The *LFKm* algorithm [9] is based on the maximization of a local metric called *subgraph fitness*, defined as follow:

$$f_{\mathcal{G}} = \frac{k_{in}^{\mathcal{G}}}{(k_{in}^{\mathcal{G}} + k_{out}^{\mathcal{G}})^{\alpha}} \tag{1}$$

In this equation $k_{in}^{\mathcal{G}}$ and $k_{out}^{\mathcal{G}}$ are respectively the in and out degree of the subgraph $\mathcal{G}$ and α is a real value used to control the size of subgraphs. Starting from this definition the authors defined the so-called *node fitness*:

$$f_{\mathcal{G}}^{A} = f_{\mathcal{G}+\{A\}} - f_{\mathcal{G}-\{A\}} \tag{2}$$

Here $f_{\mathcal{G}+\{A\}}$ and $f_{\mathcal{G}-\{A\}}$ are the fitness of subgraph $\mathcal{G}$ with and without node A. The intuitive idea is that if node A gives a high positive contribute to the subgraph fitness then it should be included in $\mathcal{G}$ and vice versa if it gives a little or negative contribute it should be excluded from $\mathcal{G}$.

To find the natural community of node A the algorithm initially define a subgraph $\mathcal{G}$ with only node A inside, then it proceeds as follow:

1. Loop over all $\mathcal{G}$ neighbours and compute the node fitness using equation 2.
2. Include in graph $\mathcal{G}$ the neighbour with the large fitness, yielding to a new subgraph $\mathcal{G}^{I}$
3. Compute the fitness of all nodes in $\mathcal{G}^{I}$ to find if someone has negative value
4. If nodes with negative fitness values were found then remove them from $\mathcal{G}^{I}$, obtaining a new subgraph $\mathcal{G}^{II}$
5. If 4 occurs continue from 3, else restart from 1 with $\mathcal{G}^{II}$

This procedure is not performed over all nodes on the network but only on a few ones chosen randomly among nodes that don't belong to any previously found subgraph. The algorithm stops when all nodes belong to at least one subgraph.

The above procedure is repeated for different α values, leading to different partitions of the original graph. The idea is that the partition which better "resists" to large changes in α values is the one that shows the strongest (most significant) community structure. To find the most stable partition the authors defined a quantitative index called *average partition fitness*:

$$\bar{f_{\mathscr{P}}} = \frac{1}{n_c} \sum_{i=1}^{n_c} f_{\mathscr{G}_i}(\alpha = 1) \tag{3}$$

Here n_c is the number of communities found inside a specific partition and $f_{\mathscr{G}_i}$ is the fitness of the subgraph $\mathscr{G}_i$ evaluated for $\alpha = 1$. If the graph has a strong community structure then the value of $\bar{f_{\mathscr{P}}}$ will not change in certain range of α, hence if we plot $\bar{f_{\mathscr{P}}}$ versus α we obtain a trend which may shows several plateaus. The most stable partition is the one obtained using the α values that produce the longest plateau in $\bar{f_{\mathscr{P}}}$ trend.

3 A Parallel Implementation on a Grid Computing Environment

The graph partition produced by *LFKm* depends strictly on the α provided for the community fitness evaluation. Finding the stable partition requires to choose a set of α values (i.e. from 0.5 to 1.5 stepping 0.01) and to run the algorithm for each of them. Every run with a different value is totally independent from the other, so it's trivial to spread the execution in a parallel environment by running n different instances of the algorithm for n different values of α. This way the total execution time is reduced to the time required for the longest instance to run, giving a big improvement to the total execution time, as shown in fig. 1.

To achieve such results we focused our attention on porting the algorithm on a grid environment, equipped with the gLite middleware [8]. The porting procedure consisted in the implementation of a bash script that act as a *wrapper* for the original algorithm, and in the definition of the job to be launched through gLite. The job description is performed using the JDL language [5] and consists in a list of attributes that set the required parameters for the parallel execution, including data requirements. In this case we need to perform a parametric job, with α as the parameter to be varied for each single instance. The wrapper is required to prepare the algorithm for the execution on the Worker Node (the computational element of the grid): it retrieves the required data, launches the algorithm for the analyse process

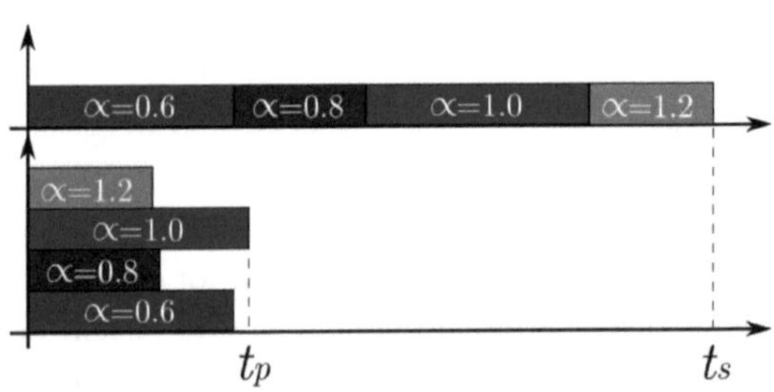

Fig. 1 A sequential execution of the algorithm compared to a parallel one

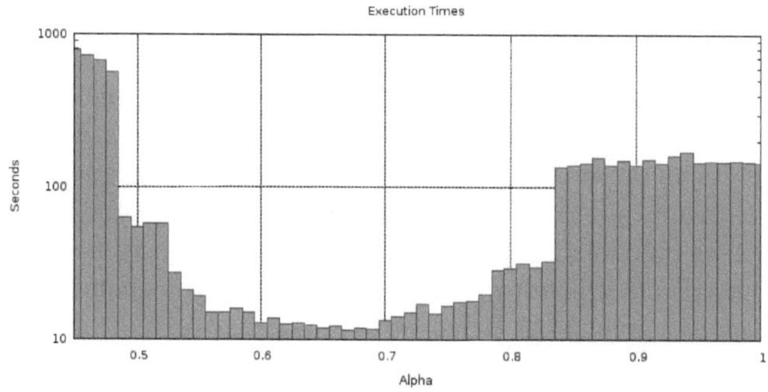

Fig. 2 Execution times for each instance associated with a specific α

and stores the computed data back to grid. To show the computational efficiency of our implementation, we tested it on a graph modeling a subset of web pages within the `.gov` domain (we call this graph `web_links`), consisting of 774,908 nodes and 4,711,340 links, and for values of α ranging from 0.45 to 1 with a step of 0.01. Using parallel execution `web_links` was analyzed in about 13 minutes, which is the longest execution time required by the job running for $\alpha = 0.45$. It is important to notice that this time rapidly decrease for bigger α: in our tests we noticed that for $\alpha = 0.5$ the execution time was less than one minute, and it keep reducing until $\alpha = 0.69$, which requires about eleven seconds (see figure 3). Then it starts growing again, reaching about 150 seconds for $0.83 \leq \alpha \leq 1$. By the way, the overall performance obtained parallelizing the whole procedure is quite outstanding: a graph with more than four millions edges was analyzed in 13 minutes only!

This analysis gives us a first trend of the $\bar{f_{\mathscr{P}}}$ function for the `web_links` graph. The trend was free of significant plateaus in the whole α range ($0.45 \leq \alpha \leq 1$), so at this point one could think that the graph has no clear community structure or the range chosen for α is inadequate. Starting from this results we proceed to compute the overlapped modularity [13] for each obtained partition; the one which better represents the real community structure will have the highest value for modularity. Using this measure instead of searching for plateaus in the $\bar{f_{\mathscr{P}}}$ trend can lead to better and more accurate results, so it can put in evidence a "meaningful" partition that is not detectable in other ways.

4 Overlapped Modularity Computation and Results

After the *LFKm* program has finished its execution, the obtained partition is ready to be processed for modularity computation. The single communities are loaded in memory one at a time and if one of them is already there it is simply discarded; this way we obtain a partition with no duplicate communities.

Subsequently, we evaluate the partition using a function that gives a measure of the quality of the community structure. In particular, in this paper we will use the overlapped modularity function as defined in [13]. This definition extends the modularity definition given by Newman in [12], introducing the possibility to take into account the effect of overlapped nodes (i.e. nodes belonging to more than one community). More specifically, given a directed graph $G(V,E)$ where V is the set of graph nodes and E is the set of links, for each graph node a *belonging* (i.e. membership) factor $\alpha_{i,k}$ expressing the strength with which node i belongs to community k is defined. Similar factors are defined for graph links by properly combining nodes belonging factors; in particular, given a link $l(i,j)$ from node i to node j, $\beta_{l,k} = \mathscr{F}(\alpha_{i,k}, \alpha_{j,k})$ expresses the strength with which link l belongs to community k. Modularity for overlapped communities is defined in [13] as:

$$Q_{ov} = \frac{1}{m} \sum_{k \in K} \sum_{i,j \in V} \left[\beta_{l(i,j),k} A_{ij} - \frac{\beta_{l(i,j),k}^{out} k_i^{out} \beta_{l(i,j),k}^{in} k_j^{in}}{m} \right] \tag{4}$$

where A_{ij} is the adjacency matrix of G, k_i^{out} and k_j^{in} are, respectively, the out–degree of node i and the in–degree of node j. In the definition of the modularity, $\beta_{l,k}$ is used to weight the contribution of link l to modularity of community k. Moreover, the terms $\beta_{l(i,j),k}^{out}$ and $\beta_{l(i,j),k}^{in}$ in Q_{ov} are tightly related to the *null–model* chosen by the authors: they represent the expected belonging factor of any possible link l, respectively, starting from and pointing to a node into community k. Notice that this formulation does not specify what kind of function $\mathscr{F}$ should be used. Following the approach of [13], the two–dimensional logistic function will be used in the experiments proposed in the present paper.

The first test we performed was about the comparison between $\bar{f_{\mathscr{P}}}$ and Q_{ov} trends for the partitions founded by the $LFKm$ algorithm with $0.73 \leq \alpha \leq 1.50$ on the Zackary network [14] (fig. 3).

The maximum Q_{ov} value (0.6963) was obtained for $1.033 \leq \alpha \leq 1.154$ which correspond to a partition of the graph in four communities. In this partition the $LFKm$ algorithm found that nodes 10, 24 and 28 are shared between two communities, so to compute Q_{ov} we fixed the belonging coefficient of each node in 0.5 that

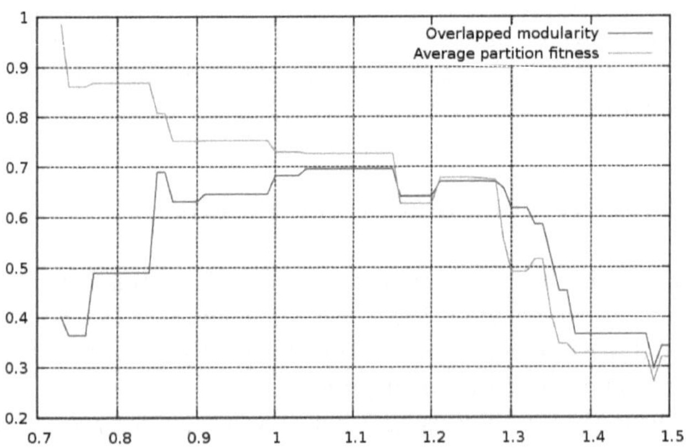

Fig. 3 Comparison between $\bar{f_{\mathscr{P}}}$ and Q_{ov} for Zachary network

means that every node belongs equally to each community it is in. This assumption is quite strong because we didn't know exactly how much a node belong to communities it is in, but we need a first approximation of this values to compute Q_{ov}. A method to found a better approximation of belonging coefficients was presented in [3]. If nodes 10, 24 and 28 are really overlapped then the Q_{ov} value should be bigger than the one found with those nodes not overlapped. To verify this we computed the overlapped modularity for a partition where nodes 10, 24 and 28 belong strictly to one community (fig. 4); for this partition we get a modularity $Q^{I}_{ov} = 0.6943$ which, as we expected, is less than Q_{ov}, where communities are overlapped.

As a last test we tried to refine the belonging coefficients of nodes 10, 24 and 28 to see if we can obtain an higher modularity value. To refine this coefficients we used the definition of *node fitness* proposed in [9]. Supposing that a node i belongs to community $k_1, k_2, \ldots, k_n$ and its fitness in each community is $f^i_{k_1}, f^i_{k_2}, \ldots, f^i_{k_n}$ we define the belonging coefficient a_{i,k_j} for node i in community k_j as:

$$a_{i,k_j} = \frac{f^i_{k_j}}{\sum_{p=1}^{n} f^i_{k_p}} \tag{5}$$

Using the belonging coefficients computed with eq. 5 we get a modularity maximum value of $Q^{II}_{ov} = 0.7003$ (see fig. 5), that is a little bigger than Q_{ov} with belonging coefficient computed as the reciprocal of number of communities a node belongs to; this result means that using eq. 5 leads to coefficients that better represent the belonging grades of nodes to communities. Another test was made on the `web_links` graph, mentioned in section 3. The analysis was made for $0.5 \leq \alpha \leq 1$ stepping 0.01. Looking at the $\bar{f_{\mathscr{P}}}$ trend obtained one can say that this graph have no stable community structure because the trend is free from significant plateaus. Looking at the Q_{ov} trend instead (fig. 6) we found a maximum value $Q_{ov} = 0.730651$ at about $\alpha = 0.62$, saying that this is the most significant partition found by the algorithm. At this value we obtained 22869 communities with an average size of about 33.8 nodes per community. Starting from this result we computed the overlapped modularity for

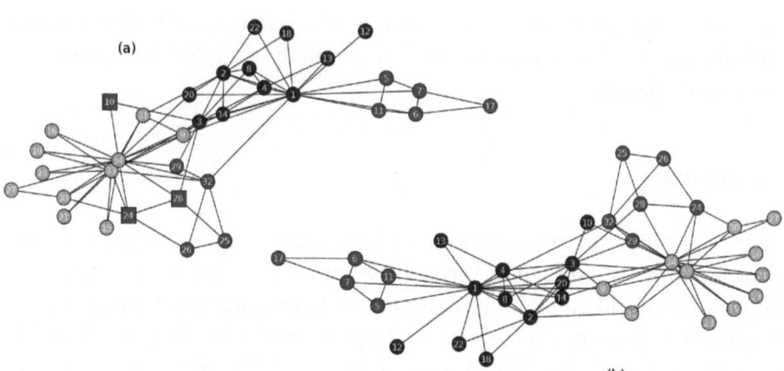

Fig. 4 Zachary partitions for $\alpha = 1.15$ with (a) overlapped and (b) non overlapped nodes

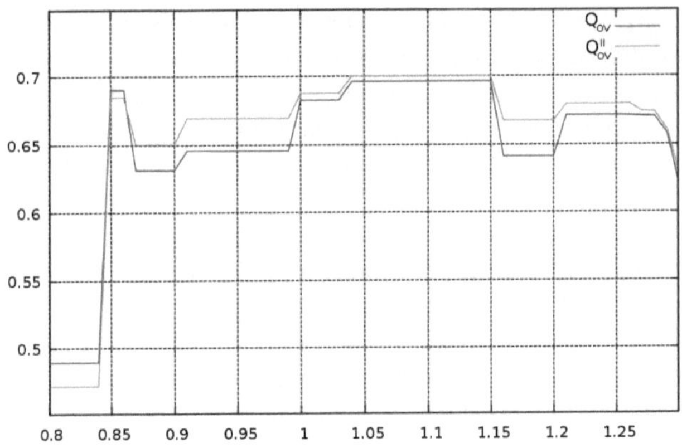

Fig. 5 Comparison between Q_{ov} and Q^{II}_{ov} trends

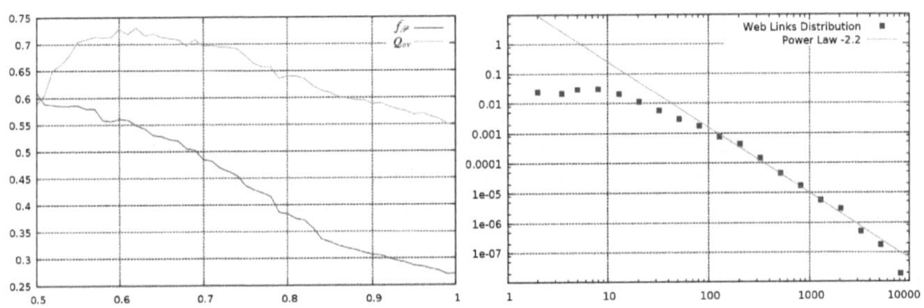

(a) (b)

Fig. 6 (a)$\bar{f_{\mathscr{P}}}$ and Q_{ov} trends for the Web Links Network and (b) its community size distribution

$\alpha = 0.62$ with refined belonging coefficients, obtaining $Q^{I}_{ov} = 0.737618$; as well as in the Zachary network the modularity with refined belonging coefficients was bigger than the one with coefficients not refined. For the partition found for $\alpha = 0.62$ we estimated the community size distribution too. We found that the tail of the distribution follow a power law with exponent -2.2 which is a quite common result for large size networks [9].

5 Conclusions

In this paper we discuss an implementation of the method for overlapped and hierarchical community structure discovering proposed in [9], showing the results of its execution within a grid computing environment. Such a high performance computing infrastructure permitted us to analyze large networks in a very short execution time, e.g. a portion of .gov WWW consisting of 774,908 nodes and 4,711,340 links, has been analyzed in 13 minutes, discovering both the overlapping communities and the hierarchical structure. Currently, only few algorithms are able to analyze networks of such large size in such a short time, thus making our approach seems both effective and efficient in addressing the communities structure discovering question.

References

1. Albert, R., Barabasi, A.-L.: Statistical mechanics of complex networks. Reviews of Modern Physics 74, 47 (2002)
2. Brandes, U., Delling, D., Gaertler, M., Gorke, R., Hoefer, M., Nikoloski, Z., Wagner, D.: On modularity clustering. IEEE Transactions on Knowledge and Data Engineering 20(2), 172–188 (2008)
3. Carchiolo, V., Malgeri, M., Mangioni, G., Nicosia, V.: Using genetic algorithms to discover overlapping communities in complex networks. In: 10th Experimental Chaos Conference (2008)
4. Danon, L., Duch, J., Diaz-Guilera, A., Arenas, A.: Comparing community structure identification (2005)
5. Job Description Language Attributes Specification for the gLite Middlewarem, http://edms.cern.ch/file/590869/1/egee-jra1-tec-590869-jdl-attributes-v0-8.pdf
6. Fortunato, S., Barthelemy, M.: Resolution limit in community detection. Proc. Natl. Acad. Sci. USA 104, 36 (2007)
7. Fortunato, S., Castellano, C.: Community Structure in Graphs. In: Encyclopedia of Complexity and System Science. Springer, Heidelberg (2008)
8. gLite Lightweight Middleware for Grid Computing, http://glite.web.cern.ch/glite/
9. Lancichinetti, A., Fortunato, S., Kertesz, J.: Detecting the overlapping and hierarchical community structure of complex networks (2008), http://www.citebase.org/abstract?id=oai:arXiv.org:0802.1218
10. Newman, M.E.J.: The structure and function of complex networks. SIAM Review 45, 167 (2003)
11. Newman, M.E.J.: Modularity and community structure in networks. Proc. Natl. Acad. Sci. USA 103, 8577 (2006)
12. Newman, M.E.J., Girvan, M.: Finding and evaluating community structure in networks. Physical Review E 69, 026113 (2004)
13. Nicosia, V., Mangioni, G., Carchiolo, V., Malgeri, M.: Extending modularity definition for directed graphs with overlapping communities (2008) arXiv:0801.1647v3
14. Zachary, W.W.: An information flow model for conflict and fission in small groups. Journal of Anthropological Research 33, 452–473 (1977)

Finding Community Structure Based on Subgraph Similarity

Biao Xiang, En-Hong Chen, and Tao Zhou

Abstract. Community identification is a long-standing challenge in the modern network science, especially for very large scale networks containing millions of nodes. In this paper, we propose a new metric to quantify the structural similarity between subgraphs, based on which an algorithm for community identification is designed. Extensive empirical results on several real networks from disparate fields has demonstrated that the present algorithm can provide the same level of reliability, measure by modularity, while takes much shorter time than the well-known fast algorithm proposed by Clauset, Newman and Moore (CNM). We further propose a hybrid algorithm that can simultaneously enhance modularity and save computational time compared with the CNM algorithm.

1 Introduction

The study of complex networks has become a common focus of many branches of science [1]. An open problem that attracts increasing attention is the identification and analysis of communities. The so-called communities can be loosely defined as distinct subsets of nodes within which they are densely connected, while sparser between which [2]. The knowledge of community structure is significant for the understanding of network evolution [3] and the dynamics taking place on networks, such as epidemic spreading [4, 5] and synchronization [6, 7].

Biao Xiang and En-Hong Chen
Department of Computer Science, University of Science and Technology of China, Hefei Anhui 230009, P.R. China
e-mail: cheneh@ustc.edu.cn

Tao Zhou
Department of Modern Physics, University of Science and Technology of China, Hefei Anhui 230026, P.R. China, and Department of Physics, University of Fribourg, Chemin du Musée 3, Fribourg 1700, Switzerland
e-mail: zhutou@ustc.edu

R. Menezes et al. (Eds.): Complex Networks: CompleNet 2009, SCI 207, pp. 73–81.
springerlink.com © Springer-Verlag Berlin Heidelberg 2009

Many algorithms for community identification have been proposed, these include the agglomerative method based on node similarity [8], divisive method via iterative removal of the edge with the highest betweenness [2, 9], divisive method based on dissimilarity index between nearest-neighboring nodes [10], a local algorithm based on edge-clustering coefficient [11], Potts model for fuzzy community detection [12], simulated annealing [13], extremal optimization [14], spectrum-based algorithm [15], iterative algorithm based on passing message [16], and so on.

Finding out the optimal division of communities, measure by modularity [9], is very hard [17], and for most cases, we can only get the near optimal division. Generally speaking, without any prior knowledge, such as the maximal community size and the number of communities, an algorithm that can give higher modularity is more time consuming [18]. As a consequence, providing accurate division of communities for a very large scale network in reasonable time is a big challenge in the modern network science. To address this issue, Newman proposed a fast greedy algorithm with time complexity $O(n^2)$ for sparse networks [19], where n denotes the number of nodes. Furthermore, Clauset, Newman, and Moore (CNM) designed an improved algorithm giving identical result but with lower computational complexity [20], as $O(n\texttt{log}^2 n)$. In this paper, based on a newly proposed metric of similarity between subgraphs, we design an agglomerative algorithm for community identification, which gives the same level of reliability but is typically hundreds of times faster than the CNM algorithm. We further propose a hybrid method that can simultaneously enhance modularity and save computational time compared with the CNM algorithm.

The rest of this paper is organized as follows. In Section 2, we introduce the present method, including the new metric of subgraph similarity and the corresponding algorithm, as well as the hybrid algorithm. In Section 3, we give a brief description of the empirical data used in this paper. The performance of our proposed algorithms for both algorithmic accuracy and computational time are presented in Section 4. Finally, we sum up this paper in Section 5.

2 Method

Considering an undirected simple network $G(V,E)$, where V is the set of nodes and E is the set of edges. The multiple edges and self-connections are not allowed. Denote $\Gamma = \{V_1, V_2, \cdots, V_h\}$ a division of G, that is, $V_i \cap V_j = \emptyset$ for $1 \leq i \neq j \leq h$ and $V_1 \cup V_2 \cup \cdots \cup V_h = V$. We here propose a new metric of similarity between two subgraphs, V_i and V_j, as:

$$s_{ij} = \frac{e_{ij} + \sum_{k=1}^{h} \frac{\sqrt{e_{ik}e_{kj}}}{|V_k|}}{\sqrt{d_i d_j}}, \tag{1}$$

where e_{ij} is the number of edges with two endpoints respectively belonging to V_i and V_j (e_{ij} is defined to be zero if $i = j$), $|V_k|$ is the number of nodes in subgraph V_k, and $d_i = \sum_{x \in V_i} k_x$ is the sum of degrees of nodes in V_i, where the degree of node x, namely k_x, is defined as the number of edges adjacent to x in $G(V,E)$. The similarity here can be considered as a measure of proximity between subgraphs, and two subgraphs having more connections or being simultaneously closely connected to some other subgraphs are supposed to have higher proximity to each other. d_i can be considered as the mass of a subgraph, and the denominator, $\sqrt{d_i d_j}$, is introduced to reduce the bias induced by the inequality of subgraph sizes. Note that, if each subgraph only contains a single node, as $V_i = \{v_i\}$, the similarity between too subgraphs, V_i and V_j, is degenerated to the well-known Salton index (also called cosine similarity in the literature) [21] between v_i and v_j if they are not directly connected.

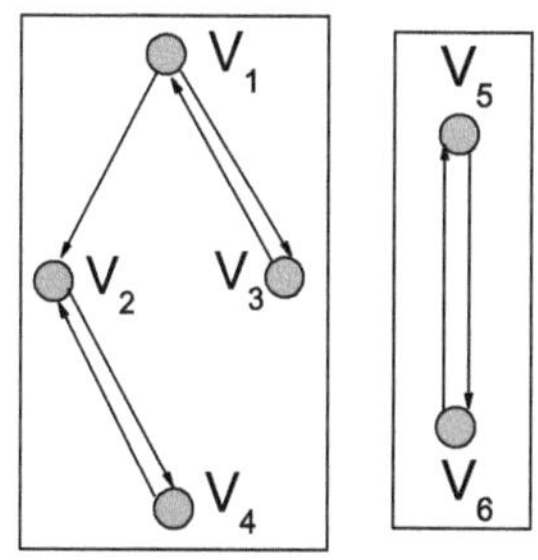

Fig. 1 Illustration of the algorithm procedure, where each node represents a subgraph. The similarities between subgraph pairs are shown in Eq. (2)

Our algorithm starts from an n-division $\Gamma_0 = \{V_1, V_2, \cdots, V_n\}$ with $V_i = \{v_i\}$ for $1 \leq i \leq n$. The procedure is as follows. (i) For each subgraph V_i, let it connect to the most similar subgraphs, namely $\{V_j | s_{ij} = \mathtt{max}_k\{s_{ik}\}\}$. (ii) Merge each connected component in the network of subgraphs generated by step (i) into one subgraph, which defines the next division. (iii) Repeat the step (i) until the number of subgraphs equals one. During this procedure, we calculate the modularity for each division and the one corresponding to the maximal modularity is recorded. To make our algorithm clear to readers, we show a small scale example consisted of six subgraphs with similarity matrix:

$$S = \begin{pmatrix} 0\,2\,2\,1\,0\,1 \\ 2\,0\,1\,3\,1\,1 \\ 2\,1\,0\,1\,0\,1 \\ 1\,3\,1\,0\,2\,0 \\ 0\,1\,0\,2\,0\,3 \\ 1\,1\,1\,0\,3\,0 \end{pmatrix}. \tag{2}$$

After the step (i), as shown in Figure 1, we get a network where each node represents a subgraph. We use the directed network representation, in which a directed arc from V_i to V_j means V_j is one of the most similar subgraphs to V_i. In the algorithmic implementation, those directed arcs can be treated as undirected (symmetry) edges. The network shown in Figure 1 is determined by the similarity matrix S, and after step (ii), the updated division contains only two subgraphs, $V_1 \cup V_2 \cup V_3 \cup V_4$ and $V_5 \cup V_6$, corresponding to the two connected components. Note that, the algorithmic procedure is deterministic and the result is therefore not sensitive to where it starts at all.

The CNM algorithm is relatively rough in the early stage, actually, it strongly tends to merge lower-degree nodes together (see Eq. (2) in Ref. [19], the first term is not distinguishable in the early stage while the enhancement of the second term favors lower-degree nodes). This tendency usually makes mistakes in the very early stage and can not be corrected afterwards. We therefore propose a hybrid algorithm which starts from a n-division $\Gamma_0 = \{V_1, V_2, \cdots, V_n\}$, and takes the procedure mentioned in the last paragraph for one round (i.e., step (i) and step (ii)). The subgraph similarity is degenerated to the similarity between two nodes:

$$s_{xy} = \frac{a_{xy} + n_{xy}}{\sqrt{k_x k_y}}, \tag{3}$$

where n_{xy} denotes the number of common neighbors between x and y, a_{xy} is 1 if x and y are directly connected, and 0 otherwise. After this round, each subgraph has at least two nodes. Then, we implement the CNM algorithm until all nodes are merged together.

3 Data

In this paper, we consider five real networks drawn from disparate fields: (i) Football.— A network of American football games between Division IA colleges during regular season Fall 2000, where nodes denote football teams and edges represent regular season games [2]. (ii) Yeast PPI.— A protein-protein interaction network where each node represents a protein [22, 23]. (iii) Cond-Mat.— A network of coauthorships between scientists posting preprints on the *Condensed Matter E-Print Archive* from Jan 1995 to March 2005 [24]. (iv) WWW.— A sampling network of the World Wide Web [25]. (v) IMDB.— Actor networks from the *Internet Movie Database* [26]. We summarize the basic information of these networks in Table 1.

Table 1 Basic information of the networks for testing

Networks	Number of Nodes, $\|V\|$	Number of Edges, $\|E\|$	References
Football	115	613	[2]
Yeast PPI	2631	7182	[22, 23]
Cond-Mat	40421	175693	[24]
WWW	325729	1090107	[25]
IMDB	1324748	3782463	[26]

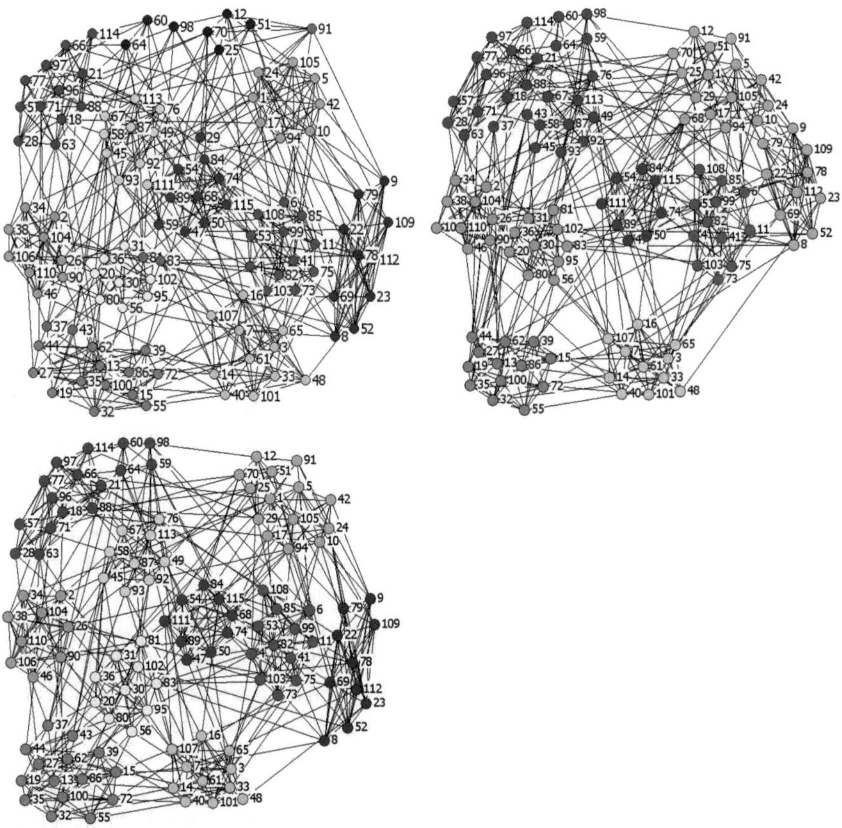

Fig. 2 Comparison of the algorithmic outputs corresponding to the best identifications subject to modularity. The three panels are (upper left panel) real grouping in regular season Fall 2000, (upper right panel) resulting communities from the CNM algorithm, and (lower panel) resulting communities from the XCZ+CNM algorithm. Each node here denotes a football team and different colors represent different groups/communities

Table 2 Maximal modularity

Algorithms	Football	Yeast PPI	Cond-Mat	WWW	IMDB
CNM	0.577	0.565	0.645	0.927	N/A
XCZ	0.538	0.566	0.682	0.882	0.691
XCZ+CNM	0.605	0.590	0.716	0.932	0.786

Table 3 CPU Time in millisecond (ms) resolution

Algorithms	Football	Yeast PPI	Cond-Mat	WWW	IMDB
CNM	172	5132	559781	12304152	N/A
XCZ	0	47	2022	17734	257875
XCZ+CNM	0	62	36422	443907	47714093

4 Result

In Table 2 and Table 3, we respectively report the maximal modularities and the CPU times corresponding to the CNM algorithm, our proposed algorithm (referred as XCZ algorithm where XCZ is the abbreviation of the authors' names), and the hybrid algorithm (referred as XCZ+CNM). All computations were carried out in a desktop computer with a single *Inter CoreE2160* processor (1.8GHz) and 2GB EMS memory. The programme code for the CNM algorithm is directly downloaded from the personal homepage of Clauset. The IMDB seems too large for the CNM algorithm, and we can not get the result in reasonable time.

From Table 2, one can find that the XCZ algorithm can provide competitively accurate division of communities verse the CNM algorithm. A significant feature of the XCZ algorithm is that it is very fast, in general more than 100 times fasters than the CNM algorithm. Just by a desktop computer, one can find out the community structure of a network containing 10^6 nodes within minutes. In comparison, the hybrid algorithm is remarkably more accurate (measured by the maximal modularity) than both the CNM and XCZ algorithms. In Figure 2, we compare the resulting community structures of the Football network, from which one can see obviously that the hybrid algorithm gives closer result to the real grouping than the CNM algorithm. We think the hybrid algorithm is fast enough for many real applications. Taking IMDB as an example, although it contains more than 1.3×10^6 nodes, the hybrid algorithm only spends less than one day. Indeed, the hybrid algorithm outperforms the CNM algorithm for both the accuracy and the speed.

5 Conclusion

Thanks to the quick development of computing power and database technology, many very large scale networks, consisted of millions or more nodes, are now available to scientific community. Analysis of such networks asks for highly efficient algorithms, where the problem of community identification has attracted more and more attentions for its hardness and practical significance.

The agglomerative method based on node similarity [8] is of lower accuracy compared with the divisive algorithms based on edge-betweenness [2] and edge-clustering coefficient [11]. In this paper, we extended the similarity measuring the structural equivalence of a pair of nodes to the so-called *subgraph similarity* that can quantify the proximity of two subsets of nodes. Accordingly, we deigned an ultrafast algorithm, which provides competitively accurate division of communities while runs typically hundreds of times faster than the well-known CNM algorithm. Using our algorithm, just by a desktop computer, one can deal with a network of millions of nodes in minutes. For example, it takes less than five minutes to get the community structure of IMDB, which is consisted of more than 1.3×10^6 nodes.

Furthermore, we integrated the CNM algorithm and our proposed algorithm and designed a hybrid method. Numerical results on representative real networks showed that this hybrid algorithm is remarkably more accurate than the CNM algorithm, and can manage a network of about one million nodes in a few hours.

The modularity has been widely accepted as a standard metric for evaluating the community identification, as well as has found some other applications such as being an assistant for extracting the hierarchical organization of complex systems [27]. Although modularity is indeed the most popular metric for community identification, and the result corresponding to the maximal modularity looks very reasonable (see, for example, Figure 2), it has an intrinsic resolution limit that makes small communities hard to detect [28, 29]. An alternative, named *normalized mutual information* [18] is a good candidate for future investigation. In addition, an extension of modularity for weighted networks, namely *weighted modularity* [30], has been adopted to deal with community identification problem in weighted networks [31, 32]. We hope the subgraph similarity proposed in this paper can also be properly extended to a weighted version to help extract the weighted communities.

Acknowledgements. This work is benefited from the *Pajek Datasets* and the *Internet Movie Database*, as well as the network data collected by Mark Newman, Albert-László Barabási and their colleagues. E.-H.C. acknowledges the National Natural Science Foundation of China under grant numbers 60573077 and 60775037. T.Z. acknowledges the National Natural Science Foundation of China under grant number 10635040.

References

1. Newman, M.E.J., Barabási, A.-L., Watts, D.J.: The Structure and Dynamics of Networks. Princeton University Press, Princeton (2006)
2. Girvan, M., Newman, M.E.J.: Community structure in social and biological networks. Proc. Natl. Acad. Sci. U.S.A. 99, 7821 (2002)
3. Palla, G., Barabási, A.-L., Vicsek, T.: Quantifying social group evolution. Nature 446, 664 (2007)
4. Liu, Z., Hu, B.: Epidemic spreading in community networks. Europhys. Lett. 72, 315 (2005)
5. Yan, G., Fu, Z.-Q., Ren, J., Wang, W.-X.: Collective synchronization induced by epidemic dynamics on complex networks with communities. Phys. Rev. E 75, 016108 (2007)
6. Arenas, A., Díaz-Guilera, A., Pérez-Vicente, C.J.: Synchronization Reveals Topological Scales in Complex Networks. Phys. Rev. Lett. 96, 0114102 (2006)
7. Zhou, T., Zhao, M., Chen, G.-R., Yan, G., Wang, B.-H.: Phase synchronization on scale-free networks with community structure. Phys. Lett. A 368, 431 (2007)

8. Breiger, R.L., Boorman, S.A., Arabie, P.: An algorithm for clustering relational data with applications to social network analysis and comparison with multidimensional scaling. J. Math. Psychol. 12, 328 (1975)
9. Newman, M.E.J., Girvan, M.: Finding and evaluating community structure in networks. Phys. Rev. E 69, 026113 (2004)
10. Zhou, H.: Distance, dissimilarity index, and network community structure. Phys. Rev. E 67, 061901 (2003)
11. Radicchi, F., Castellano, C., Ceccon, F., Loreto, V., Parisi, D.: Defining and identifying communities in networks. Proc. Natl. Acad. Sci. U.S.A 101, 2658 (2004)
12. Reichardt, J., Bornholdt, S.: Detecting Fuzzy Community Structures in Complex Networks with a Potts Model. Phys. Rev. Lett. 93, 218701 (2004)
13. Guimerà, R., Sales, M., Amaral, L.A.N.: Modularity from fluctuations in random graphs and complex networks. Phys. Rev. E 70, 025101 (2004)
14. Duch, J., Arenas, A.: Community detection in complex networks using extremal optimization. Phys. Rev. E 72, 027104 (2005)
15. Newman, M.E.J.: Finding community strcuture in networks using the eigenvectors of matrics. Phys. Rev. E 74, 036104 (2006)
16. Frey, B.J., Dueck, D.: Clustering by Passing Messages Between Data Points. Science 315, 972 (2007)
17. Brandes, U., Delling, D., Gaertler, M., Görke, R., Hoefer, M., Nikoloski, Z., Wagner, D.: On Finding Graph Clusterings with Maximum Modularity. In: Brandstädt, A., Kratsch, D., Müller, H. (eds.) WG 2007. LNCS, vol. 4769, pp. 121–132. Springer, Heidelberg (2007)
18. Danon, L., Díaz-Guilera, A., Duch, J., Arenas, A.: Comparing community structure identification. J. Stat. Mech., P09008 (2005)
19. Newman, M.E.J.: Fast algorithm for detecting community strcuture in networks. Phys. Rev. E 69, 066133 (2004)
20. Clauset, A., Newman, M.E.J., Moore, C.: Finding community structure in very large networks. Phys. Rev. E 70, 066111 (2004)
21. Salton, G., McGill, M.J.: Introduction to Modern Information Retrieval. McGraw-Hill, Auckland (1983)
22. von Merging, C., Krause, R., Snel, B., Cornell, M., Oliver, S.G., Fields, S., Bork, P.: Comparative assessment of large-scale data sets of protein-protein interactions. Nature 417, 399 (2002)
23. Bu, D., Zhao, Y., Cai, L., Xue, H., Zhu, X., Lu, H., Zhang, J., Sun, S., Ling, L., Zhang, N., Li, G., Chen, R.: Topological structure analysis of the protein-protein interaction network in budding yeast. Nucleic Acids Research 31, 2443 (2003)
24. Newman, M.E.J.: The structure of scientific collaboration networks. Proc. Natl. Acad. Sci. U.S.A. 98, 404 (2001)
25. Albert, R., Jeong, H., Barabási, A.-L.: Diameter of the World Wide Web. Nature 401, 130 (1999)
26. Ahmen, A., Batagelj, V., Fu, X., Hong, S.-H., Merrick, D., Mrvar, A.: Visualisation and Analysis of the Internet Movie Database. In: Proceedings of the 2007 Asia-Pacific Symposium on Visualization, pp. 17–24. IEEE Press, Los Alamitos (2007)
27. Sales-Pardo, M., Guimerà, R., Moreira, A.A., Amaral, L.A.N.: Extracting the hierarchical organization of complex systems. Proc. Natl. Acad. Sci. U.S.A. 104, 15224 (2007)
28. Fortunato, S., Bartheémy, M.: Resolution limit in community detection. Proc. Natl. Acad. Sci. U.S.A. 104, 36 (2007)

29. Lancichinetti, A., Fortunato, S., Radicchi, F.: Benchmark graphs for testing community detection algorithms. Phys. Rev. E 78, 046110 (2008)
30. Newman, M.E.J.: Analysis of weighted networks. Phys. Rev. E 70, 056131 (2004)
31. Fan, Y., Li, M., Zhang, P., Wu, J., Di, Z.: The effect of weight on community structure of networks. Physica A 378, 583 (2007)
32. Mitrović, M., Tadić, B.: Search of Weighted Subgraphs on Complex Networks with Maximum Likelihood Methods. In: Bubak, M., van Albada, G.D., Dongarra, J., Sloot, P.M.A. (eds.) ICCS 2008, Part II. LNCS, vol. 5102, pp. 551–558. Springer, Heidelberg (2008)

Structural Trends in Network Ensembles*

Ulrik Brandes, Jürgen Lerner, Uwe Nagel, and Bobo Nick

Abstract. A collection of networks is considered a network ensemble if its members originate from a common natural or technical process such as repeated measurements, replication and mutation, or massive parallelism, possibly under varying conditions. We propose a spectral approach to identify structural trends, i. e. prevalent patterns of connectivity, in an ensemble by delineating classes of networks with similar role structure. Formal, experimental, and practical evidence of its potential is given.

1 Introduction

Network-analytic studies most frequently are concerned with a small set of networks if not a singleton instance. Indicators employed in such analyses range from properties of individual actors (e. g., centrality and role) and local patterns (e. g., reciprocated ties, stars, and closed triangles) over to global network characteristics (e. g., density, modularity, and degree distributions) [17, 4]. Given the ever increasing availability of data, there is a growing tendency to compare families of networks that, e.g., may be defined on different sets of actors or encode different relations (see, e. g., [10, 7, 9]). Application scenarios for network comparison include examination whether different teams of employees exhibit structural differences [8, p. 81], comparison of networks among different species [9], detection of user-roles in Usenet newsgroups by patterns in egocentric reply-networks [18], and comparison of

Ulrik Brandes, Jürgen Lerner, Uwe Nagel, and Bobo Nick
Department of Computer & Information Science, University of Konstanz
e-mail: `Ulrik.Brandes@uni-konstanz.de`

* Research supported in part by DFG under grant GK 1024 (Research Training Group "Explorative Analysis and Visualization of Large Information Spaces") and University of Konstanz under grant FP 626/08.

R. Menezes et al. (Eds.): Complex Networks: CompleNet 2009, SCI 207, pp. 83–97.
springerlink.com © Springer-Verlag Berlin Heidelberg 2009

social integration in personal networks of immigrants [6]. To emphasize the (assumed) existence of an inherent relation in a collection of networks, we will refer to it as a *network ensemble.*

Clearly, the elements of a network ensemble can be compared and categorized based on any global structural property or extrinsic attributes (i. e., "who is in the network"). In this paper we treat networks as similar, if they exhibit the same *role structure*, i. e., if they show the same pattern of connectivity among classes of actors. Actors are said to play the same *role*, or occupy the same *position*, in a network if they are similarly connected to other actors that themselves play the same role [3, 17, 13]. For instance, by this definition university professors would occupy the same structural position if they have identical patterns of ties to students, secretaries, industry contacts, other professors and so on. Such a role assignment on a given network yields a smaller graph, called the *role graph* (compare [14]), whose vertices are the actor classes and whose (weighted) edges encode how actors in one class are on average connected to actors in the other class. In this paper we compare networks by the role graphs they give rise to. Returning to the above example, the networks of two universities might differ in the fact that in one university the professors are differently (stronger or weaker) connected to the students than in the other university.

However, since already the decision problem whether a given graph admits a role structure of a specific type is NP-complete [11], our strategy for network comparison seems to run into serious computational problems. Indeed we do not attempt to design an algorithm that is able to compare any (worst-case) instance of a network ensemble; rather, we propose an efficient heuristic algorithm that provably performs well on networks ensembles arising from certain random graph models. More specifically, if a network ensemble contains subsets of networks that indeed differ sufficiently in their role structure, then our algorithm will correctly distinguish those networks with high probability, i. e., it will detect a good clustering of the network ensemble.

In Sect. 2 we define a stochastic model for network ensembles with latent role structures and define the associated clustering problem. We propose an algorithm for clustering network ensembles in Sect. 3 and show in Sect. 4 that it recovers class-memberships with high probability—given that the stochastic model satisfies certain preconditions. Experimental results on artificially generated networks in Sect. 5 and a small case study on Wikipedia edit-networks in Sect. 6 provide further evidence of the usefulness of our approach.

2 A Network Ensemble Model with Latent Roles

We start by recalling a model for random graphs that exhibit a hidden (latent) class structure; such a model is defined, e. g., in [15] and [12].

Definition 1. A *planted partition model* $\mathcal{G}(n, k, \psi, P)$ is given by a number of vertices n, a number of classes k, a partition $\psi\colon \{1, \ldots, n\} \to \{1, \ldots, k\}$ of the n vertices into k classes and a symmetric $k \times k$ matrix P of edge probabilities $P_{ij} \in [0, 1]$ between classes. The probability of a given graph $G = (V, E)$ with n vertices given the model $\mathcal{G}(n, k, \psi, P)$ is

$$\mathbb{P}\left(G \middle| \mathcal{G}(n, k, \psi, P)\right) = \prod_{\{u,v\} \in E} P_{\psi(u)\psi(v)} \prod_{\{u,v\} \notin E} 1 - P_{\psi(u)\psi(v)}$$

Alternatively, an instance G of $\mathcal{G}(n, k, \psi, P)$ is drawn by including each edge $\{u, v\}$ into G independently with probability $P_{\psi(u)\psi(v)}$. Thus, the probability of an edge between vertices u and v is only dependent on their class-membership.

A planted partition model $\mathcal{G} = \mathcal{G}(n, k, \psi, P)$ is completely defined by its *expected adjacency matrix* which is the $n \times n$ matrix $\overline{M} = \overline{M}(\mathcal{G})$ whose entries are defined by $\overline{M}_{ij} = P_{\psi(i)\psi(j)}$. Note that $\overline{M}$ is indeed the expectation of the adjacency matrices of graphs drawn from $\mathcal{G}(n, k, \psi, P)$.

In this paper we consider random network ensembles that are mixtures of such planted partition models.

Definition 2. A *(planted partition) network ensemble* $\mathcal{E}(N, K, \Psi, \mathcal{G}_1, \ldots, \mathcal{G}_K)$ is given by a number of graphs N, a number of graph models K, an assignment $\Psi\colon \{1, \ldots, N\} \to \{1, \ldots, K\}$ of the N graphs to the K models and a family of K planted partition models $\mathcal{G}_1, \ldots, \mathcal{G}_K$, where $\mathcal{G}_i = \mathcal{G}(n_i, k_i, \psi_i, P_{(i)})$.

Thus, a planted partition network ensemble is a set of random graphs drawn from planted partition models. To obtain an instance of $\mathcal{E}(N, K, \Psi, \mathcal{G}_1, \ldots, \mathcal{G}_K)$, the N graphs G_i, $i = 1, \ldots, N$, are independently drawn from the planted partition model $\mathcal{G}_{\Psi(j)}$. For sake of simplicity we will often write in this paper network ensemble instead of planted partition network ensemble. In the major part of this paper we consider network ensembles with the same number of vertices; only in Sect. 6, where we analyze real-world networks, do we apply our algorithms to networks of different size.

The *algorithmic problem* associated with a planted partition network ensemble $\mathcal{E} = \mathcal{E}(N, K, \Psi, \mathcal{G}_1, \ldots, \mathcal{G}_K)$ is the following.

> Given an instance $(G_1, \ldots, G_N)$ of $\mathcal{E}$, classify the N graphs such that two graphs are in the same class if and only if they are drawn from the same underlying planted partition model.

Obviously, without any further preconditions this problem is not solvable. (For instance, if two of the underlying planted partition models are identical, the graphs generated from these are not distinguishable.) However, in this paper we propose an efficient algorithm such that, given certain preconditions, we can decide for each given pair of graphs with high probability whether they are drawn from the same underlying model or not. (The term *with high probability* means "with probability that tends to one as the size of the graphs tends to infinity"; this notion is often employed to assess the quality of heuristic algorithms, compare [15].)

3 Classification Method

3.1 Intuition

Let $(G_1, \ldots, G_N)$ be an instance drawn from a planted partition network ensemble $\mathcal{E}(N, K, \Psi, \mathcal{G}_1, \ldots, \mathcal{G}_K)$. In the following we sketch how we will proceed to determine for any two graphs whether they are drawn from the same underlying planted partition model or not.

A very simple observation is that if we were not given the adjacency matrices $M_1, \ldots, M_N$ of $G_1, \ldots, G_N$ but rather their expectation values $\overline{M_1}, \ldots, \overline{M_N}$, then the problem would be fairly trivial: under the minimal assumption that the planted partition models $\mathcal{G}_1, \ldots, \mathcal{G}_K$ are pairwise different, it follows that their expected adjacency matrices are pairwise different as well. Hence, two graphs G_i, G_j out of $G_1, \ldots, G_N$ are drawn from the same model if and only if their expected adjacency matrices $\overline{M_i}$, $\overline{M_j}$ are equal.

However, our algorithm does not have access to the expected adjacency matrices. Indeed, the adjacency matrix M_i of an instance graph is rather very far from its expectation value $\overline{M_i}$. (Note that M_i is a zero/one-matrix, while the entries of $\overline{M_i}$ are from the real interval $[0, 1]$; thus the expectation is typically not attainable.)

What helps us out of this dilemma is a well-known combination of results from matrix perturbation theory [16] with probabilistic bounds on the eigenvalues of random matrices [1] (also compare [15]). Basically, these results enable us to show that, even if the adjacency matrix M of an instance graph differs entrywise very much from its expectation $\overline{M}$, the spectrum of M is with high probability close to the spectrum of $\overline{M}$. It follows that the adjacency matrices of two graphs drawn from the same model have (with high probability) similar spectra and, under the assumption that the spectra of the expected adjacency matrices differ in at least one value, graphs from different models have a larger difference in their spectra.

3.2 Method

The ordered spectrum of a symmetric $n \times n$ matrix is denoted by $\lambda_1 \leq \ldots \leq \lambda_n$ and the vector $\lambda(M) = (\lambda_1, \ldots, \lambda_n)^T$ is referred to as the spectrum vector of matrix M. An instance of our classification problem is created by randomized drawing N adjacency matrices M_i according to some underlying role graphs. Each adjacency matrix M_i provides us with a corresponding graph G_i, which gives us a network ensemble $\mathcal{E} = \{G_1, \ldots, G_N\}$.

We do neither know how many role graphs there are nor which graphs belong to the same role graph. What we do know is that graphs being drawn from the same role graph should have a spectrum much more similar to each other than graphs drawn from different role graphs. As we show in Sec. 4 it is suggestive to measure the similarity between two graphs in this context by the supremum norm of their spectrum vectors. So under certain assumptions $\|\lambda(M_1) - \lambda(M_2)\|_\infty$ should be much greater if the graphs corresponding to M_1 and M_2 are created from different role graphs than if they were from the same role graph.

This makes our classification problem to a classical clustering problem. Given objects and distances between them, dense clusters of objects are searched. Standard clustering algorithms can be applied as long as they can be parameterized with a distance measure. An example would be a version of k-means, which does not need the number of clusters as an input. We performed some promising experiments with an iterated k-means using the silhouette coefficient to decide for the optimal clustering.

The pseudocode in Alg. 1 summarizes our method for detecting structural trends in network ensembles.

Algorithm 1. Structural Trends in Network Ensembles

Input: network ensemble $\mathcal{E} = \{G_1, \ldots, G_N\}$
Result: clustering $\{C_1, \ldots, C_k\}$ with $\mathcal{E} = \biguplus_i C_i$

for $G \in \mathcal{E}$ **do**
| determine spectrum vector $\lambda(G)$
end
partition $\{\lambda(G) : G \in \mathcal{E}\}$ using supremum norm

In the ideal case this method extracts from an arbitrary ensemble a classification of the graphs into groups having the same role graph and thereby solves the stated algorithmic problem. Taking our results one could also think of classifications of ensembles consisting of differently sized graphs. It would be necessary to restrict the spectrum vector to a size such that it can be determined for all graphs of the ensemble. One would also have to take care of the growth of the eigenvalues which is linear in the number of vertices of the graph. A possible approach is to take the n eigenvalues with maximum absolute value of each graph where n is the size of the smallest graph in the given instance and divide them by the size of the graph. A more efficient method could be inferred by knowledge of the sizes of the underlying role graphs. If the assumptions of the next section are met, the number of eigenvalues used can be limited by the maximum number of vertices of the role graphs without changing the defined distances.

3.3 Generalization to Weighted Networks

We restricted the method sketched above to binary (unweighted) graphs only for notational simplification. A model for ensembles of *weighted* networks (i. e., graphs with real edge-weights) could be defined in almost the same way as in Sect. 2. A *weighted* planted partition model is defined as in Def. 1 with the difference that when drawing an instance graph one does not include (unweighted) edges with a given probability but rather the weight of an edge $\{u, v\}$ is drawn from a distribution dependent on the classes of u and v. Examples of such distributions would include the normal distribution where the mean value depends on the vertex classes.

The adjacency matrix of a weighted graph is a real matrix whose entries encode the edge weights. Note that the abovementioned method for network classification via the eigenvalues of graphs can be applied to these weighted matrices without any change. Furthermore, the theorems that will be presented in Sect. 4 hold also for the case of weighted matrices. The application to real world data sketched in Sect. 6 indeed analyzes an ensemble of weighted networks.

4 Evidence from Matrix Perturbation Theory

Let $(G_1, \ldots, G_N)$ be an instance drawn from a planted partition network ensemble $\mathcal{E}(N, K, \Psi, \mathcal{G}_1, \ldots, \mathcal{G}_K)$ whose underlying graph models have a common number of vertices n. Building on results from matrix perturbation theory, we show in this section that for sufficiently large n (and ignoring a small number of outliers) the spectra of graphs drawn from the same model have smaller distance than the spectra of graphs drawn from different models.

We start by associating a planted partition model $\mathcal{G}$ with a matrix $A(\mathcal{G})$ that encodes the relative class-sizes as well as the edge-probabilities between classes of $\mathcal{G}$. It turns out that the eigenvalues of $A(\mathcal{G})$ correspond—up to a multiplicative constant that is related to the size of the classes—to the non-zero eigenvalues of the expected adjacency matrix $\overline{M}(\mathcal{G})$.

Definition 3. Let $\mathcal{G} = \mathcal{G}(n, k, \psi, P)$ be a planted partition model and denote the proportion of vertices in class $i = 1, \ldots, k$ with

$$q_i = |\{v;\ 1 \le v \le n \text{ and } \psi(v) = i\}| \,/n$$

The *structure matrix* associated to $\mathcal{G}$ is the $k \times k$ matrix $A = A(\mathcal{G})$ whose entries are defined by $A_{ij} = \sqrt{q_i q_j} \cdot P_{ij}$.

To make the notion *with high probability* precise, we define a process by which we can increase the number of vertices in a planted partition model without changing its structure (more precisely: without changing the relative class-sizes nor the edge-probabilities between classes). Let $\mathcal{G}_1 = \mathcal{G}(n_1, k, \psi_1, P)$ be a fixed planted partition model and $t \in \mathbb{N}_{\ge 1}$ an integer. A planted partition model $\mathcal{G}_t$ that has $n_t = t \cdot n_1$ vertices and the same structure matrix as $\mathcal{G}_1$ can be defined by $\mathcal{G}_t = \mathcal{G}(n_t, k, \psi_t, P)$, where $\psi_t \colon \{1, \ldots, n_t\} \to \{1, \ldots, k\}$ with $\psi_t(v) = \psi_1(\lceil v/t \rceil)$. Note that it holds $A(\mathcal{G}_t) = A(\mathcal{G}_1)$.

The next theorem shows that the eigenvalues of a planted partition model with fixed structure matrix grow linearly in the number of vertices.

Theorem 1. *Let $\mathcal{G}_1 = \mathcal{G}(n_1, k, \psi_1, P)$ be a planted partition model, $t \in \mathbb{N}_{\geq 1}$ an integer, and set $n_t = t \cdot n_1$. Each eigenvalue λ of $A(\mathcal{G}_1)$ yields an eigenvalue $n_t \cdot \lambda$ of $\overline{M}(\mathcal{G}_t)$. The remaining $n_t - k$ eigenvalues of $\overline{M}(\mathcal{G}_t)$ are equal to zero.*

Proof. Note first that the expected matrix $\overline{M} = \overline{M}(\mathcal{G}_t)$ is (after reordering the vertices such that vertices in the same class are consecutive) an $n_t \times n_t$ block matrix

$$\overline{M} = \begin{pmatrix} B_{11} & \ldots & B_{1k} \\ \vdots & & \vdots \\ B_{k1} & \ldots & B_{kk} \end{pmatrix} \text{ with blocks } B_{ij} = \begin{pmatrix} P_{ij} & \ldots & P_{ij} \\ \vdots & & \vdots \\ P_{ij} & \ldots & P_{ij} \end{pmatrix}$$

of dimension $(q_i \cdot n_t) \times (q_j \cdot n_t)$. (Note that $q_i \cdot n_t$ is indeed an integer which follows from the definitions of q_i and n_t.)

Let $x = (x_1, \ldots, x_k)^{\mathrm{T}}$ be any eigenvector of $A(\mathcal{G}_1)$ associated to eigenvalue $\lambda \in \mathbb{R}$. Spelling out the equation $A(\mathcal{G}_1) \cdot x = \lambda \cdot x$ yields for $i = 1, \ldots, k$

$$\lambda \cdot x_i = \sum_{j=1}^{k} \sqrt{q_i q_j} P_{ij} x_j = \sqrt{q_i} \sum_{j=1}^{k} \sqrt{q_j} P_{ij} x_j \ . \tag{1}$$

We claim that the n_t-dimensional vector y defined by

$$y = (\underbrace{x_1/\sqrt{q_1}, \ldots, x_1/\sqrt{q_1}}_{n_t \cdot q_1 \text{ times}}, \ldots, \underbrace{x_k/\sqrt{q_k}, \ldots, x_k/\sqrt{q_k}}_{n_t \cdot q_k \text{ times}})^{\mathrm{T}} \ ,$$

satisfies $\overline{M}(\mathcal{G}_t) \cdot y = n_t \lambda y$ which shows that $n_t \lambda$ is an eigenvalue of $\overline{M}(\mathcal{G}_t)$ and, thus, yields the assertion of the theorem.

To see that this is true let v be any integer satisfying $1 \leq v \leq n_t$ and let $i = \psi_t(v)$ (i. e., i is the index of the class of vertex v.) We get

$$(\overline{M}(\mathcal{G}_t) \cdot y)_v = \sum_{j=1}^{k} n_t q_j P_{ij} x_j / \sqrt{q_j} = n_t \sum_{j=1}^{k} \sqrt{q_j} P_{ij} x_j = n_t \lambda x_i / \sqrt{q_i} = n_t \lambda y_v$$

where the third equation follows from Eq. (1). □

Corollary 1. *Let $\mathcal{G}_1$ and $\mathcal{H}_1$ be two planted partition models with the same number of vertices n. Let $t \in \mathbb{N}_{\geq 1}$ and set $n_t = t \cdot n$. Under the assumption that the eigenvalues of $A(\mathcal{G}_1)$ and $A(\mathcal{H}_1)$ differ in at least one value, the distance between the spectrum vectors of the expected adjacency matrices of $\mathcal{G}_t$ and $\mathcal{H}_t$ grows linearly in the number of vertices n_t. More precisely*

$$\|\lambda(\overline{M}(\mathcal{G}_t)) - \lambda(\overline{M}(\mathcal{H}_t))\|_\infty = n_t \cdot \|\lambda(A(\mathcal{G}_1)) - \lambda(A(\mathcal{H}_1))\|_\infty \in \Theta(n_t) \ .$$

All that remains us to do is to bound the difference between the eigenvalues of the adjacency matrix M of an instance graph and its expectation $\overline{M}$. For this purpose define the perturbation matrix $E = M - \overline{M}$ as the difference between the observed adjacency matrix and its expectation. We recall a result from matrix perturbation theory.

Theorem 2 ([16]). *Let $M = \overline{M} + E$ be a symmetric perturbation of a symmetric matrix $\overline{M}$. Then we have*

$$\|\lambda(M) - \lambda(\overline{M})\|_\infty \leq \|E\|_2 \ ,$$

where $\|E\|_2$ denotes the maximal absolute value of an eigenvalue of E.

The second result we need is a probabilistic bound on the maximal eigenvalue of the difference between the observed adjacency matrix and its expectation.

Theorem 3 ([15]). *Let M, $\overline{M}$ and E be defined as above and let n denote their dimension. Let σ^2 be the largest variance of an entry in M. (Note that if the i, j'th entry of $\overline{M}$ equals p, then its variance is $p - p^2$; the variance is non-zero if p is in the open interval from zero to one.) If $\sigma^2 \gg \log^6 n/n$, then $\|E\|_2 \leq 4\sigma\sqrt{n}$ with probability at least $1 - 2e^{-\sigma^2 n/8}$.*

The assumption $\sigma^2 \gg \log^6 n/n$ is satisfied for sufficiently large n if at least one entry of $\overline{M}$ is different from zero and one. For the remainder of this paper we will take this assumption for granted; note that this excludes only uninteresting cases.

The next corollary follows from Theorems 2 and 3.

Corollary 2. *Let M and $\overline{M}$ be defined as above and let n denote their dimension. It is $\|\lambda(M) - \lambda(\overline{M})\|_\infty \in \mathcal{O}(\sqrt{n})$ with probability in $1 - o(1)$ (i. e., with probability tending to one as n tends to infinity).*

Combining these results enables us to show the following result which indicates that any reasonable clustering on the spectrumvectors will—apart from a small proportion of outliers— correctly assign the networks into clusters according to the underlying graph models.

Theorem 4. *Let $\mathcal{E} = \mathcal{E}(N, K, \Psi, \mathcal{G}_1, \ldots, \mathcal{G}_K)$ be a network ensemble in which the underlying graph models have a common number of vertices n_t. For each $\varepsilon > 0$ there exists $n_0 \in \mathbb{N}$ such that for $n_t \geq n_0$ we have for any instance of $\mathcal{E}$*

$$\|\lambda(G) - \lambda(G')\|_\infty < \varepsilon \cdot \|\lambda(H) - \lambda(H')\|_\infty$$

for any graphs G and G' drawn from the same model and any graphs H and H' drawn from different models, with probability in $1 - o(1)$.

Proof. The following assertions hold with high probability. By Corollary 2 it is $\|\lambda(G) - \lambda(G')\|_\infty \in \mathcal{O}(\sqrt{n_t})$. Let $\overline{M}$ be the expected adjacency matrix of H and $\overline{M'}$ be the expected adjacency matrix of H'. By Corollary 1 it is $\|\lambda(\overline{M}) - \lambda(\overline{M'})\|_\infty \in \Theta(n_t)$ and, again by Corollary 2 we have $\|\lambda(H) - \lambda(\overline{M})\|_\infty \in \mathcal{O}(\sqrt{n_t})$ and $\|\lambda(H') - \lambda(\overline{M'})\|_\infty \in \mathcal{O}(\sqrt{n_t})$. Together if follows $\|\lambda(H) - \lambda(H')\|_\infty \in \Theta(n_t)$ which implies that for sufficiently large n_t the inequality of the theorem is satisfied. □

Note that Theorem 4 makes only assertions for specific numbers of vertices of the form $n = t \cdot n_1$. However, this restriction is only necessary for notational simplification. In the next section we provide evidence by simulation that the spectra of the different graph clusters are well separated for all sufficiently large values of n. Furthermore this simulation indicates which values of n are sufficiently large for the theorems to hold.

5 Experimental Evidence

To estimate the tightness and expandability of our results we conducted experiments on artificially generated ensembles. Experiments on these examples split up in two major categories. The first is the case where we specify some simple role graphs on two nodes and try to determine the size needed to distinguish graphs drawn from these models by their spectra. For the second part of the study, role graphs were generated from random edge distributions and random group sizes. In all experiments graphs of different sizes were generated from each model and compared pairwise in terms of the $\| \ \|_\infty$ norm on their spectrum vectors. In the choice of graph sizes we did not restrict ourself to cases that make an exact matching of the group sizes possible but we also integrated graphs where group sizes can only be approximately established.

Although our analytical results apply only to exact matches of the group sizes, our experiments suggest that our method can be used, e.g., in a setting in which the group membership of each node is determined randomly from a distribution where the probability for membership in a class equals the relative class size in the model. This method was used in the experiment on prespecified models and in the second part of the experiment on random models. Here additionally the case of group sizes matched as exact as possible is examined.

The outcome of our experiments is a diagram showing distances between graphs of these examples. Here we distinguish the distances between graphs drawn from different models (points in grey) and those from the same model (in black). What we expect is that distances between graphs from different models grow faster than distances between graphs from identical models with growing graph sizes. The diagrams show the development of distances between graphs in ensembles for growing graph sizes. The size of the graphs

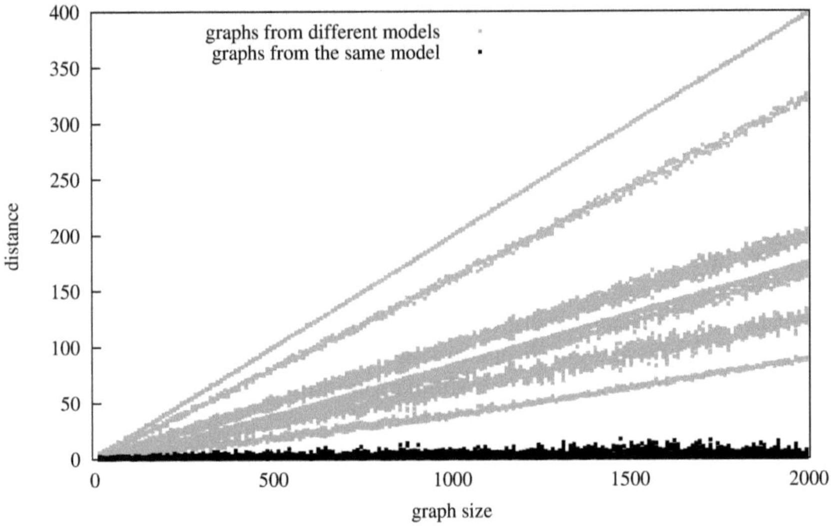

Fig. 1 Pairwise distances in ensembles generated for six prespecified role graphs with two nodes each. For ensembles with graphs of 300 vertices, a simple distance threshold separates classes well

is shown on the horizontal axis and the distance between the graphs appears on the vertical axis.

5.1 Prespecified Role Graphs

For illustrative purposes we start with some archetypical partition models that have been selected for their simplicity and good separation. Two edge probabilities $p = 0.2$ and $q = 10^{-3}$ are used and every possible symmetric edge distribution for a two-node graph on these values is generated. Excluding isomorphic role graphs this yields six edge distributions for us to cluster. For this experiment we chose a uniform class distribution and while generating instances from our models we do not try to match class sizes as exact as possible but rather assign vertices to classes uniformly at random. Ensembles consisting of graphs with size $10i$ where examined for $i = 2, \ldots, 100$. The main result can be seen in Fig. 1, which suggests that a clear separation by spectrum vectors should be possible for graphs having about 300 vertices. For random 200-node graphs we give three samples for each of the six models in Fig. 2; structural trends are clearly recognizable.

5.2 Random Role Graphs

For the next experiments we generated five role graphs with seven nodes each. For each partition model the desired class sizes and the edge distribution were

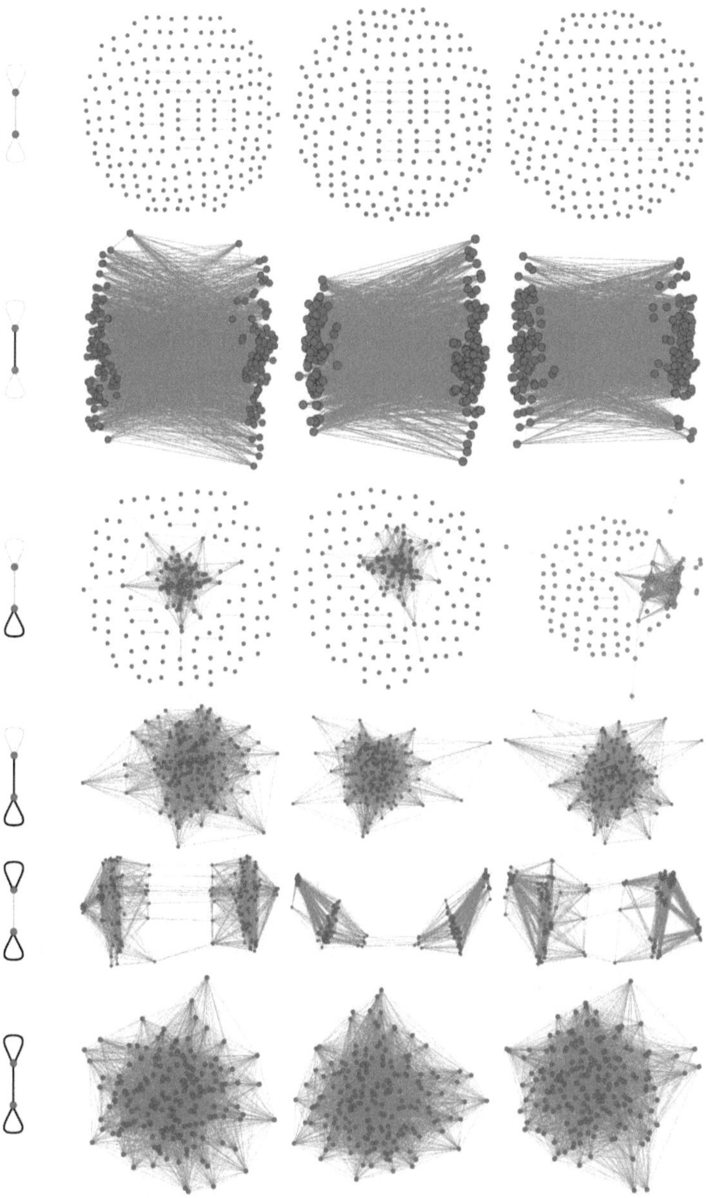

Fig. 2 Sample graphs with 200 vertices each from six prespecified two-node role graphs

drawn randomly and independent from a uniform distribution over [0, 1]. Basically each role graph consists of two random matrices, a $n \times 1$ matrix for node distribution to classes and a $n \times n$ matrix for the edge distribution. The

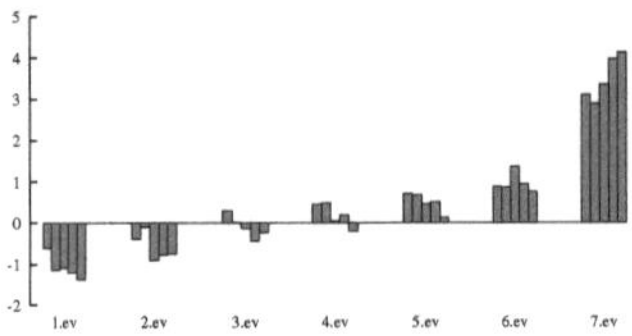

(a) spectra

model	2	3	4	5
1	0.528	0.509	0.859	1.022
2		0.810	1.089	1.252
3			0.619	0.782
4				0.388

(b) distances

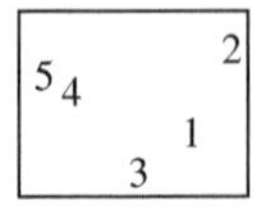

(c) plot

Fig. 3 Spectra and pairwise distances of five randomly selected role graphs

graphs drawn from the different models are distinguished by their spectrum vectors and those again derive from the corresponding model.

As can be seen from the comparison in Fig. 3, the corresponding models do not differ much. A table with pairwise distances in supremum norm and an overview of these distances in a two dimensional layout obtained via multidimensional scaling quantify their relative shapelessness.

Since there are no pointed differences in these spectra, the sampled models can be considered quite typical. In particular, they form a classification instance much harder than the prespecified models used in the previous section. This is supported by additional experiments on different role graphs created in the same way and giving similar results, but not reported here.

For $i = 2, \ldots, 200$ ensembles were created consisting of five graphs with $10i$ vertices for each model, which gives us a sample ensemble with 25 graphs for every i. The difference in the two experiments lies in the assignment of vertices to classes. While in the first part it was tried to match partition sizes as exact as possible, in the second part the approach described above was used where desired partition sizes are used as a distribution.

Figure 4 shows how the distances between graphs drawn from different role graphs diverge from each other such that 10 different rays of dots can be seen, which is expected when the distances between the role graphs differ pairwise. Consider a graph with a node for each role graph and edge weights defined by the distance between adjacent role graphs measured as described above. The edge weights of this graph growing linearly in the number of vertices of the graphs the ensemble contains plus some random noise are the rays that can be seen in the diagram. The bottom line in black consists of distances between graphs drawn from the same role graph.

As an unexpected result the distances between graphs corresponding to the same role graph seem to be constant which could be a hint that the established borders are not tight.

The diagram in Fig. 5 shows how the divergence is weakened by inexact partition sizes. Compared to Fig. 4 a clear distinction between graphs drawn from different and those drawn from equal role graphs is achieved only with graphs having significantly more vertices, even though a trend towards clear separation can be observed.

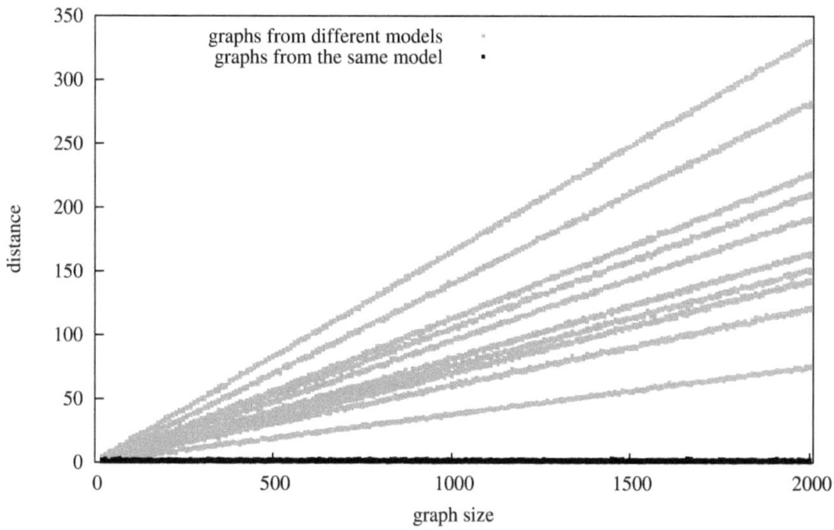

Fig. 4 Distance development with class sizes matched as accurate as possible

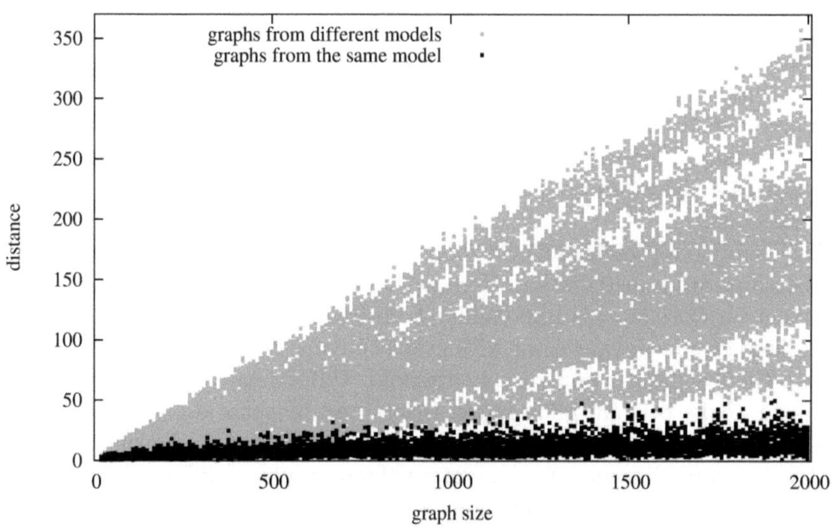

Fig. 5 Distance development with class sizes as distribution

6 Practical Evidence

In this section we want to demonstrate the performance of our graph distance in an application on real world data. We analyzed the edit networks of Wikipedia articles (see [5] for the definition of edit networks) and retrieved the expected result that average distances between networks with a supposed common structure are smaller than those with an expected difference in structure. For the analysis we randomly chose 60 articles with at least 1000 edits and 60 networks that were labeled 'featured' by the Wikipedia community.

From the edit logs of these articles a complete graph with a node for each author was created. Each edge was weighted by sums of negative edits between the adjacent authors. A negative edit occurs if either one author deletes words written by the other or if he restores words that were deleted by the other and is valued by the logarithm of the number of words deleted/restored. Since the edit graphs have in general different sizes we had to restrict the comparison to graphs having at least 500 vertices and vectors consisting of the 500 eigenvalues with biggest absolute value divided by the number of vertices. The number 500 was chosen since the differences are not expressed that clear with smaller values. For greater values the number of graphs being left is not meaningful for class comparison since noise and outliers could dominate the results.

The distance between classes was computed as the average of the pairwise distances between all graphs of the corresponding classes, while the distances between two graphs was measured as the above described distance on the spectrumvector of their weighted adjacency matrices.

The computations yield average distances of $21.7 \cdot 10^{-3}$ within the arbitrary chosen articles, $15.7 \cdot 10^{-3}$ in the class of featured articles and an average distance of $20.9 \cdot 10^{-3}$ between the two classes. As expected the featured articles tend toward a structure in their edit graphs that is common among this class and distinguishable from those of arbitrary articles. The fact that the inner class difference of arbitrary articles is higher than the distance to the featured articles can be easily explained by the fact that featured articles are a subclass.

This example represents an even more general case than the one where class memberships are a distribution. Here we have differently sized graphs and a statement on class sizes is impossible. Additionally we are not dealing with unweighted graphs anymore but with graphs having weighted edges. We tried to drop the weights by applying a threshold. Unfortunately in this scenario too much of the original information is lost and no separation between the classes can be seen at all. This drove us to use our method beyond proved effectiveness, on adjacency matrices of weighted graphs of different sizes. The obtained results support our decisions and encourage further examination of possible applications in this direction.

7 Conclusion

We introduced a spectral approach to identify groups of networks with similar role structure, i. e., networks that show the same pattern of connectivity among actor-classes, in network ensembles. We provided evidence for the usefulness of this method by probabilistic arguments (Sect. 4), by simulation results (Sect. 5), and by analyzing an ensemble of empirical networks generated from the edit-history of sampled Wikipedia articles (Sect. 6). In previous work, network ensembles have often been described by other indicators such as density, degree sequences, or nearest neighbor connectivity (see e. g., [2]). Note that such approaches are not in competition, but orthogonal to our method, since they are based on different assumptions about the underlying ordering principle.

References

1. Alon, N., Krivelevich, M., Vu, V.H.: On the concentration of eigenvalues of random symmetric matrices. Israel Journal of Mathematics 131(1), 259–267 (2002)
2. Bianconi, G.: The entropy of randomized network ensembles. Europhysics Letters 81, 28005 (2008)
3. Borgatti, S.P., Everett, M.G.: Notions of position in social network analysis. Sociological Methodology 22, 1–35 (1992)
4. Brandes, U., Erlebach, T. (eds.): Network Analysis. Springer, Heidelberg (2005)
5. Brandes, U., Kenis, P., Lerner, J., van Raaij, D.: Network analysis of collaboration structure in Wikipedia. In: Proc. 18th Intl. World Wide Web Conf. (WWW 2009) (to appear, 2009)
6. Brandes, U., Lerner, J., Lubbers, M.J., McCarty, C., Molina, J.L.: Visual statistics for collections of clustered graphs. In: Proc. IEEE Pacific Visualization Symp (PacificVis 2008), pp. 47–54 (2008)
7. Butts, C.T., Carley, K.M.: Some simple algorithms for structural comparison. Computational & Mathematical Organization Theory 11(4), 291–305 (2005)
8. Carley, K.M., Lee, J.-S., Krackhardt, D.: Destabilizing networks. Connections 24(3), 79–92 (2002)
9. Faust, K.: Comparing social networks: Size, density, and local structure. Metodološki zvezki 3(2), 185–216 (2006)
10. Faust, K., Skvoretz, J.: Comparing networks across space and time, size and species. Sociological Methodology 32(1), 267–299 (2002)
11. Fiala, J., Paulusma, D.: The computational complexity of the role assignment problem. In: Baeten, J.C.M., Lenstra, J.K., Parrow, J., Woeginger, G.J. (eds.) ICALP 2003. LNCS, vol. 2719, pp. 817–828. Springer, Heidelberg (2003)
12. Golub, B., Jackson, M.O.: How homophily affects communication in networks (2008), `http://arxiv.org/abs/0811.4013`
13. Leicht, E.A., Holme, P., Newman, M.E.J.: Vertex similarity in networks. Physical Review E 73 (2006)
14. Lerner, J.: Role assignments. In: Brandes and Erlebach [4], pp. 216–252
15. McSherry, F.: Spectral partitioning of random graphs. In: Proceedings of the 42nd Annual IEEE Symposium on Foundations of Computer Science (FOCS 2001), pp. 529–537 (2001)
16. Stewart, G.W., Sun, J.-G.: Matrix Perturbation Theory. Academic Press, London (1990)
17. Wasserman, S., Faust, K.: Social Network Analysis: Methods and Applications. Cambridge University Press, Cambridge (1994)
18. Welser, H.T., Gleave, E., Fisher, D., Smith, M.: Visualizing the signatures of social roles in online discussion groups. Journal of Social Structure 8 (2007)

Generalized Attachment Models for the Genesis of Graphs with High Clustering Coefficient

Jens Gustedt

Abstract. Commonly used techniques for the random generation of graphs such as those of Erdős & Rényi and Barabási & Albert have two disadvantages, namely their lack of bias with respect to history of the evolution of the graph, and their incapability to produce families of graphs with non-vanishing prescribed clustering coefficient. In this work we propose a model for the genesis of graphs that tackles these two issues. When translated into random generation procedures it generalizes the above mentioned procedures. When just seen as composition schemes for graphs they generalize the perfect elimination schemes of chordal graphs. The model iteratively adds so-called *contexts* that introduce an explicit dependency to the previous evolution of the graph. Thereby they reflect a historical bias during this evolution that goes beyond the simple degree constraint of preference edge attachment. Fixing certain simple statical quantities during the genesis leads to families of random graphs with a clustering coefficient that can be bounded away from zero.

Keywords: attachment models, random graph generation, k-trees.

1 Introduction and Overview

Modeling the genesis of graphs has become an important task in many application domains, most prominent are probably graphs of computer networks and social networks. Such models are an important prerequisite for the random generation of realistic large networks that are needed for simulations in the framework of the application domains and also for testing graph algorithms and their implementations on examples of a realistic size.

Application graphs that follow a vein of 'families of hidden cliques defining a graph' have been identified by Guillaume and Latapy [2004]: protein networks, the

Jens Gustedt
INRIA Nancy – Grand Est, Villers-lés-Nancy, France
e-mail: `jens.gustedt@loria.fr`

R. Menezes et al. (Eds.): Complex Networks: CompleNet 2009, SCI 207, pp. 99–113.
springerlink.com © Springer-Verlag Berlin Heidelberg 2009

core network of the Internet, web connections (http links), the co-starring relation among film actors and the co-occurrence relation of words in sentences. The goal of this paper here is to give a random model that covers the inter-relationship between those cliques, *e.g* the non-trivial overlap of those cliques, and the temporal evolution of these networks respectively their genesis.

During the genesis of such graphs, the dependence of the process from previous choices is an important detail that we try to handle. We propose a relatively simple model, in which each newly introduced clique depends on a previously known one. For the graph of co-authorships, e.g, a new paper often emerges from a previous one by slightly modifying the list of authors, some people cease contributing for the new one, others, such as experts of a particular subdomain or new PhD students join in.

Classical random graph models, such as promoted by Erdős and Rényi [1960], usually do not fulfill the necessities from the application domains since the expected structure of the generated graphs is too far from what is observed in practical settings. Because of that, starting form the work of Barabási and Albert [1999] in the last decade a lot of attempts to provide more realistic models have been undertaken, see *e.g* Latapy [2007] for an overview, or Dorogovtsev and Mendes [2003] for a textbook.

These models try to capture different statistical properties of the generated graphs, such as the degree distribution or the expected distance between arbitrary pairs of vertices. Here, we will concentrate on an important property that is desired for random graphs, namely their density: it was observed that real world graphs are generally sparse (‘*have much less edges than they could*’) but are usually locally quite dense (‘*the probability that two neighbors of a vertex are also connected is high*’).

The *clustering coefficient* $\overline{cc}(G)$ of a graph G is meant to capture this feature. It measures how close or how far the vertices are to being simplicial: a vertex v is called *simplicial* if the neighborhood of v is a clique. If v is not simplicial, the *neighborhood density* $cc(G, v)$ at v is the quotient between the number of edges inside the neighborhood of v and the maximum number of such edges. Formally $\overline{cc}(G)$ is given as the average over all vertices $v \in V$ of

$$cc(G, v) = \begin{cases} 1 & \text{if } v \text{ is simplicial} \\ \frac{|E_v G|}{\binom{deg_G(v)}{2}} & \text{otherwise} \end{cases} \tag{1}$$

Observe that for the (practically unimportant) border case of vertices of degree 0 or 1 this definition is different from what can be usually found in the literature, see *e.g* Dorogovtsev and Mendes [2003], but it will prove convenient in the sequel. Using only the quotient would not be well defined since the denominator then would be 0 for these special cases. With our particular choice we have that $\overline{cc}$ will be high for trees: for any tree G without vertices of degree 2, $\overline{cc}(G) > 0.5$. This high value fits well to what we obtain when we investigate other graphs that show a ‘tree-like’ structure.

A k-tree, see Arnborg [1985], Robertson and Seymour [1986], is a graph that can be obtained from a clique of size $k+1$ by iteratively joining new vertices to cliques of size k. The so-called k-tree-decomposition of such a graph is easily defined from this iterative definition.

Observation 1. *For some $k > 0$ let G be a k-tree with a k-tree-decomposition that has no vertices of degree 2. Then $\overline{cc}(G) \geq \frac{1}{2} - \frac{k-2}{2|V(G)|}$.*

A proof for this statement can be obtained by counting the leaves of the decomposition: they correspond to simplicial vertices in the graph. Because of the restriction for the tree-decomposition at least half of the nodes of that decomposition are leaves and thus almost one half of the vertices of the graph are simplicial. The observation then follows by straightforward estimations.

The bound given in Observation 1 is not sharp. It accounts only for those vertices that are in fact simplicial and the impact of the other vertices is neglected. But on the other hand it shows us some reason why some graph G might have a high clustering coefficient namely that it might have '*a lot*' (here one half) of '*boundary*' vertices that are simplicial.

Also, Observation 1 puts a restriction on the permissible k-tree decomposition that G should have. It is easy to see that for the path with n edges $\overline{cc}(P_n) = 2/n$. So the bound or a similar one can't hold without that restriction and for $k = 1$. But other vertices, if not simplicial, still might contribute with a high value of cc. Since the number of potential edges in the neighborhood grows quadratically with the size of that neighborhood low degree vertices will in fact easier fulfill such a condition. One aim of this paper is to make such an observation into a precise counting argument which will prove that in some classes of sparse graphs there will always be enough low degree vertices that contribute with high values to the clustering coefficient.

To fulfill our goal of randomly generating graphs that have a high clustering coefficient we will generalize the constructions that lead to k-trees and chordal graphs. Our construction uses so-called contexts that are analogous to the cliques of a tree-decomposition. Section 2 introduces our model in detail. Section 2 then shows the relationship to previously known graph classes and proves some basic properties of the construction.

In Section 3 we will prove a more general bound for the clustering coefficient that is obtained by this construction. By this we can guarantee that all random graphs that are generated by it will have a non-vanishing clustering coefficient. In contrast to Observation 1, this will also take non-simplicial vertices into account.

To formulate the full statement of the bound as it is given there is not possible without the notations that are introduced in Section 2. But translated to the special case of k-trees it reads as follows.

Lemma 1. *Let G be a k-tree for some $k > 1$. Then $\overline{cc}(G) \geq \dfrac{k-1}{4k^2}$.*

Observe, that compared to Observation 1 here we got rid of the restriction on the shape of the tree-decomposition. The lower bound on k is necessary because of examples like P_n as mentioned above. If a graph contains an abundant number of induced P_2 its clustering coefficient can be brought arbitrarily low.

2 The Model and Its Basic Properties

We will attempt to model the genesis of large interconnection networks. As mentioned above, our model will be an extension of the construction that leads to k-trees and more generally to chordal graphs. The idea of this paper is that we will distinguish the *observable*, generally a graph of relations, from an implicit family of *concepts* or *contexts* that define it, but which are in general not or only partially observable. These contexts correspond to the cliques of the tree-decomposition of chordal graphs. The genesis of these structures will be described as a *process*, *i.e* as an evolution of a combinatorial object *in time*.

This idea that the edges of the graph under investigation come from a more or less hidden structure of cliques was already implicit in Ravasz and Barabási [2003] and has been verified for a large number of application graphs, see Guillaume and Latapy [2004]. Implicitly it also is present in [Dorogovtsev and Mendes, 2003, Sec. 5.13 ff.] where techniques for the growth of graphs are introduced and for which a bound of $\overline{cc}$ is given. In fact, in our terminology this technique boils down to the generation of k-trees and what will be described in the sequel can be viewed as a generalization of that approach. As a basic example throughout this paper we will use an interconnection network of which we, as scientists, are all concerned: the graph of co-authorship. In that graph the vertex set is formed by the 'objects' Ob under investigation which in the example in fact are 'subjects', namely the authors that have been contributing to a specific scientific domain. We add an edge $e = \{ob_1, ob_2\} \in E$ if ob_1 and ob_2 have co-authored a scientific paper of the domain.

As we already see in this basic example, the implicit structure that we investigate is richer than the just the graph (Ob, E). In particular we have an important family of implicit objects or co-objects Co which are the scientific papers. Each such paper $co \in Co$ describes the context of a collaboration between a set of colleagues and the relational structure E is derived from them.

Other structures that follow the same vein of 'families of hidden cliques defining a graph' can be found in application graphs:

- For metabolic networks bio-chemical reactions define cliques.
- For the Internet on IP level a direct IP connection between two hosts is usually provided through a connection to the same switch or router, or by listening to a shared physical medium (radio or ethernet). Thus the Internet graph can be seen as generated from the local 'link layer' networks that each form local cliques.
- For social networks, connection between individuals can often be attributed to the common membership of in a social structure such as the family, the work place, school classes, church etc.
- For semantic networks two concepts (scientific papers, web pages, ...) are linked if they are co-referenced in some text (other paper, other web page, ...). Here the referring context defines a clique of all its referred objects.

What the modeling in previous papers failed to explain (up to our best knowledge) was the inter-relationship between those cliques, *e.g* the non-trivial overlap of those cliques, and the temporal evolution of these networks respectively their genesis.

τ	-2	-1	0	1	2	3	4
ob_0	0			0	0		0
ob_1	1						
ob_2	2			2			
ob_3		3	3			3	
ob_4		4					
ob_5		5	5				
ob_6			6	6	6	6	
ob_7					7		7
ob_8						8	8

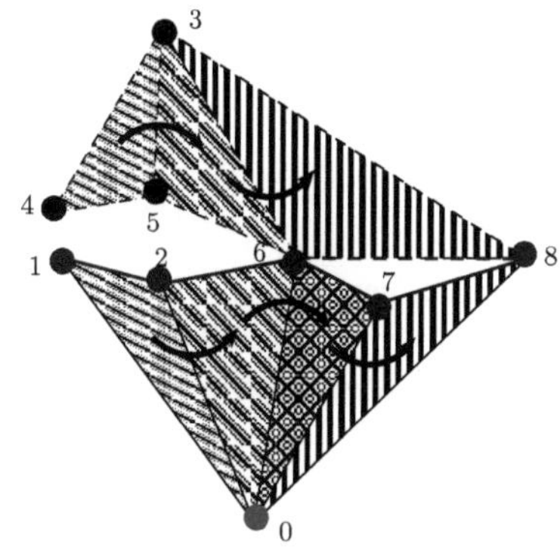

Fig. 1 A graph that is generated by a sequence of contexts. On the left, columns corresponds to contexts and are given in generation order. Lines denote the contexts in which an object appears. The boxed entries denote the initial subsequence where a new element is '*created*'. On the right, the contexts correspond to triangles. The edges of the graph are the edges of these triangles

Objects and Their Contexts. More formally, we will investigate pairs (Ob, Co) where Ob is a set (usually finite) and $Co \subseteq 2^{Ob}$ is some family of subsets over Ob. We will refer to Ob as the set of *objects*, *e.g* the members of a scientific community, and to Co as *contexts* in which these objects appear together, *e.g* the scientific papers that they co-author. We will say that $ob_1, ob_2 \in Ob$ are *linked* if there is some $co \in Co$ such that $\{ob_1, ob_2\} \subseteq co$. The set of edges, relations or links E is then defined by $E_{Ob,Co} = \{\{ob_1, ob_2\} \mid \exists co \in Co, ob_1 \text{ and } ob_2 \in co \text{ with } ob_1 \neq ob_2\}$.

First, observe that from that definition (Ob, E) has no loops. Second, in this formal definition Ob and Co play 'opposite' sites in a bipartite relation that is defined by the containment relation $\in$. In view of the combinatorial structure the emphasis of Ob being the first set of the pair and Co the second is arbitrary ("just" given by the application). For the example, we could equally well be interested in the relationship among the papers, linking two papers if they share a common author. A context for papers then corresponds to the oeuvre of a scientist.

These pairs (Ob, Co) are considered as parts of a process (the 'genesis') of a growing structure. Namely we look at sequences $((Ob_\tau, Co_\tau))_{\tau=(-\aleph+1)\ldots,0,1,\ldots}$ where the parameter τ can be thought of as discretized time, $\aleph > 0$ defines a number of pre-existent contexts, and we have that

$$Ob_{-\aleph+1} \cdots \subseteq Ob_0 \subseteq Ob_1 \subseteq \cdots \quad \text{and} \quad Co_{-\aleph+1} \cdots \subseteq Co_0 \subseteq Co_1 \subseteq \cdots$$

In terms of the link graph this defines a growing sequence of graphs $G_i = (Ob_i, E_i)$, with $E_i = E_{Ob_i, Co_i}$:

$$(\emptyset, \emptyset) = (Ob_{-\aleph}, E_{-\aleph}) \subset (Ob_{-\aleph+1}, E_{-\aleph+1}) \cdots \subseteq (Ob_0, E_0) \subseteq (Ob_1, E_1) \subseteq \cdots$$

Figure 1 shows an example of a sequence of contexts and the resulting graphs.

What is usually observed in applications is only part of the genesis, *e.g* some or just one of the graphs. The number of vertices (resp. edges) at time τ are denoted with n_τ and m_τ respectively, *i.e*

$$\mathrm{n}_\tau = |Ob_\tau| \qquad \text{and} \qquad \mathrm{m}_\tau = |E_\tau|\,.$$

To describe such a genesis we will assume that one step from (Ob_τ, Co_τ) to $(Ob_{\tau+1}, Co_{\tau+1})$ is given by exactly one new context. That is, there is an enumeration of the contexts $\ldots, co_0, co_1, co_2, \ldots$ such that

$$Ob_\tau = \bigcup_{t\leq\tau} co_t \qquad \text{and} \qquad Co_\tau = \bigcup_{t\leq\tau} \{co_t\}.$$

The potentially infinite base sets for $\tau \to \infty$ are denoted Ob_∞ and Co_∞. Generally, we will also suppose that the sequence has no *redundancy*, *i.e* that for all τ there are $ob, ob' \in Ob_\tau$ such that $\{ob, ob'\} \notin E_{\tau-1}$. For all τ we will denote this set of non-redundant edges $\bar{E}_\tau = E_\tau \setminus E_{\tau-1}$ for which we thus have $\bar{E}_\tau \neq \emptyset$. Another property that we assume for the sequence is that it respects inclusion in the following sense. For $\tau < \kappa$, the new elements that appear in Ob_τ fulfill

$$Ob_\kappa \setminus Ob_{\tau-1} \not\subset Ob_\tau \setminus Ob_{\tau-1}, \tag{2}$$

i.e no context appearing later than co_τ in the sequence will add less elements to $Ob_{\tau-1}$ than co_τ.

Even with this property the exact ordering of the contexts will be arbitrary. In fact, if $Ob_\tau = Ob_{\tau+1}$ the contexts co_τ and $co_{\tau+1}$ can be considered *interchangeable*. A subsequence $co_\tau, \ldots, co_{\tau+\ell}$ in $(co_i)_{i=0,\ldots}$ is *stable* if all adjacent elements are interchangeable, or, in other words $Ob_\tau = \cdots = Ob_{\tau+\ell}$. It is *maximally stable* if it is stable and may not be extended to the left or right without loosing that property. We then also have that $Ob_\tau \cap Ob_{\tau-1} = \cdots = Ob_{\tau+\ell} \cap Ob_{\tau-1}$.

With that definition we may subdivide our sequence uniquely into maximally stable subsequences. For each τ, $start_\tau$ denotes the start index of the maximal stable subsequence to which co_τ belongs. $\mathcal{L}_\tau$ denotes the number of contexts in that subsequence. Both values are independent of the particular ordering of the subsequence. Also we associate to each such maximal stable subsequence the set of newly introduced objects, $create_\tau = Ob_\tau \setminus Ob_{start_\tau - 1}$.

The Starting Point of the Genesis. In a genesis as we attempt to describe here, new objects and contexts will emerge from ones that previously exist. Clearly this is only possible if we assume the existence of some of them initially. In a sequence of contexts we will thus assume that there is a finite number $\aleph$ of predefined contexts $co_{-(\aleph-1)}, \ldots, co_0$. The parameters n_0 and m_0 are thus the number of vertices and edges that we assume present before the genesis starts, and which we assume to be finite numbers.

Besides some more or less obvious requirements (*e.g* that we only may connect to a clique of a required size if there is one), the statistical properties of the graphs that will result below will not much depend on the initial choice of contexts. They will be dominated by the many other choices during the genesis. Such as the cristallization germ of a snowflake is very important for it to form initially but by itself it will not have much influence on the final shape.

The Bias Introduced by Imitation. In our genesis, the dependence of the process from previous choices is an important detail that we have to handle. We propose a relatively simple model, in which each new $co \in Co$ depends on one previously known other element. In our example of the graph of authorship, a new work often emerges from a previous one by slightly modifying the list of authors, some people cease contributing for the new one, others, such as experts of a particular subdomain or new PhD students join in.

So in general we suppose that for each $\tau > 0$ there is $\rho(\tau) < \tau$, such that for

$$stab_\tau = co_{\rho(\tau)} \cap co_\tau \qquad old_\tau = co_{\rho(\tau)} \setminus co_\tau \qquad new_\tau = co_\tau \setminus co_{\rho(\tau)} \tag{3}$$

we have that $stab_\tau \neq \emptyset$ and $new_\tau \neq \emptyset$. For constructing co_τ a pre-existing $co_{\rho(\tau)}$, the *paragon*, is chosen, copied into a new set co_τ in which old_τ is replaced by new_τ.

Now, the type of transformations that are permitted when going from $co_{\rho(\tau)}$ to co_τ will be much dependent on the particular domain; different sets of rules will lead to specific families of graphs. We will investigate simple deterministic and statistical properties of such rules. Therefore, we will introduce some parameters on the sizes of these sets that could describe the evolution in different application domains, either by following some deterministic rule, or just by some statistical correlation. These parameters may then be used to describe an observed sequence or to randomly sample a 'typical' member of a specific family.

- $\mathcal{D}_\tau$ the *size* of co_τ
- $\mathcal{O}_\tau$ the number of replaced objects, $|old_\tau|$
- $\mathcal{N}_\tau$ the number of replacing objects, $|new_\tau|$
- $\mathcal{S}_\tau$ the number of *sporadic* objects that had been present before this stable subsequence but are reintroduced into the new context, *i.e* $|new_\tau \setminus create_\tau|$.
- $\mathcal{L}_\tau$ the length of the maximal stable sequence containing co_τ as defined above

We expect that applications usually could provide sensible values (resp. distributions) for these parameters. In the case of the co-authorship graph, e.g, it should be possible to describe the distribution of the number of authors of a paper ($\mathcal{D}$), the typical number of papers to which a young PhD student contributes at the beginning of his career ($\mathcal{L}$), the scientific heritage from the initial environment (relation ρ).

Special Cases for Fixed Parameters. In the following we will look at cases such that the parameters from the previous section are equal to some constant for all $\tau > 0$. A sequence which fulfills these constraints for fixed values of $\mathcal{D}, \mathcal{O}, \mathcal{N}, \mathcal{S}$, and $\mathcal{L}$ will be called a $(\mathcal{D}, \mathcal{O}, \mathcal{N}, \mathcal{S}, \mathcal{L})$*-sequence*. Whenever we fix only some of these parameters we will replace those that are not fixed by the symbol '*'.

Fixing the values for some of these parameters lead to the genesis of families of graphs that are already well studied. Here we always assume that the initial start of the genesis is always given by just one initial context co_0. A schema for the inter-relationship of such graph families is given in Figure 2. In particular, known classes correspond to the following values:

$(\mathcal{D}, \mathcal{O}, \mathcal{N}, \mathcal{S}, \mathcal{L})$

$(2, 1, 1, 0, t)$ The contexts themselves are just edges, one of the objects of the paragon is replaced by the new object and every such new object appears as new in a stable sequence of length exactly t for some constant $t > 0$. This leads to the genesis by preferential edge attachment as described by Barabási and Albert [1999]. Suppose that for each τ the choice of ρ_τ is uniform among all possible values. Then, each vertex v is chosen with a probability that is proportional to its degree at instance τ.

$(2, 2, 2, 1, *)$ This is the limit case for which a complete renewal for each new edge that is added is performed. It leads to random graphs similar to the model of Erdős and Rényi [1960]; at any point choosing any of the pairs (v, w) is equally likely.

$(*, *, *, 0, 1)$ Each new context introduces new vertices and doesn't push old vertices into new contexts. This leads to the genesis of chordal graphs. It is easy to see that our genesis is a reverted elimination ordering, that ρ defines a tree and that that tree is a tree decomposition of the graph. In the restricted case that $\mathcal{D}_\tau \equiv k + 1$ is also fixed and $\mathcal{O}_\tau \equiv 1$, *i.e* $(k + 1, 1, 1, 0, 1)$-sequences, the graphs are the k-trees. For $\mathcal{D}_\tau = 2$, *i.e* $(2, 1, 1, 0, 1)$, we then just have trees.

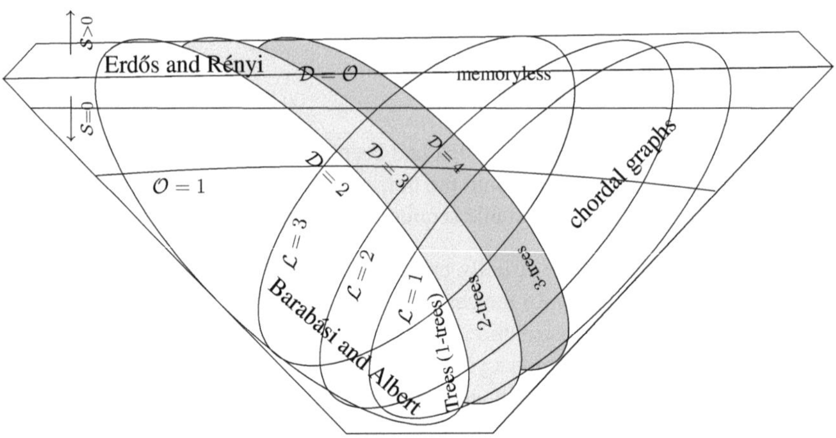

Fig. 2 Special cases of graph classes in function of the parameters $\mathcal{D}$, $\mathcal{O}$, $\mathcal{S}$ and $\mathcal{L}$. The topline '*memoryless*' refers to the fact that in the case that $\mathcal{D} = \mathcal{O}$ (*e.g* Erdős and Rényi graphs) each choice is independent from previous choices. The blob with $\mathcal{L} = 1$ are the chordal graphs, and intersecting this blob with those for $\mathcal{D} = 2, 3, \ldots$ gives rise to k-trees

3 Parameter Estimations

Estimating the Number of Edges. An important feature of most real world graphs is their sparseness, *i.e* the fact that they usually have an average degree that is bounded by some small constant. In this section we will show how such a claim holds for our proposed genesis. In the next section, such bounds then will be a major ingredient to prove a bound on the clustering coefficient.

An important case for bounding the clustering coefficient from below will be sequences that don't have sporadic occurences of objects, *i.e* with $\mathcal{S} = 0$. As seen above these occur as generalizations of well studied classes of graphs, so studying the resulting families might be of interest of its own.

Lemma 2. *Let $\mathcal{S} = 0$ and $\tau = start_\tau$, i.e τ is the starting point of a maximal stable sequence. Then the number of edges $|E_\tau \setminus E_{\tau-1}|$ added by context co_τ is $|stab_\tau| \cdot |create_\tau| + \binom{|create_\tau|}{2}$.*

Proof. Let $stab_\tau = co_\tau \cap co_{\rho(\tau)} = \{ob_1, \dots, ob_d\}$ be the objects that had been copied into co_τ. Since all of them are already present in $co_{\rho(\tau)}$, no new edge between them is added at time τ.

Now let $ob \in create_\tau$ be any new object. By definition all edges induced for it in co_τ must be new. So we get $|stab_\tau|$ new edges for it that link to older elements.

Among the objects in $create_\tau$ we create all pairs of possible edges. □

Let us now restrict even further by fixing $\mathcal{O} = 1$ and $\mathcal{S} = 0$, *i.e* classes where going from $co_{\rho(\tau)}$ to co_τ replaces *exactly one* object $ob \in old_\tau$ by the unique element $ob' \in create_\tau$.

For the simplicity of the arguments we will first assume that the following property holds for ρ: if $\tau < \tau'$ are such that co_τ and $co_{\tau'}$ are members of the same stable sequence ($create_\tau = create_{\tau'}$) then their predecessors by ρ are mutually disjoint.

$$co_{\rho(\tau)} \cap co_{\rho(\tau')} = \emptyset \tag{4}$$

We call such a sequence *distinctive*.

Lemma 3. *For a distinctive sequence with $\mathcal{O} = 1$ and $\mathcal{S} = 0$, the number of edges $|E_\tau \setminus E_{\tau-1}|$ added by context co_τ is $\mathcal{D}_\tau - 1$.*

Proof. If $\tau = start_\tau$, *i.e* is first in its stable subsequence, this is Lemma 2, since $\mathcal{O} = |create_\tau| = 1$, $|stab_\tau| = |co_\tau| - 1$ and thus $|stab_\tau| \cdot |create_\tau| = |co_\tau| - 1$.

If $\tau \neq start_\tau$, we have that $co_h \cap co_\tau = create_\tau$ for all $start_\tau \leq h < \tau$. Thus no edge that is induced by co_τ may have been created previously. □

Corollary 1. *For a distinctive sequence with $\mathcal{O} = 1$ and $\mathcal{S} = 0$, we have that*

$$|E_\tau| = m_0 - \tau + \textstyle\sum_{\theta \leq \tau} \mathcal{D}_\theta = m_0 + \tau(\widetilde{\mathcal{D}}_\tau - 1). \tag{5}$$

Lemma 4. *Suppose that for a distinctive sequence with $\mathcal{O} = 1$ and $\mathcal{S} = 0$ there are constants δ^+ and λ^+ with $\widetilde{\mathcal{D}}_\tau \leq \delta^+$ and $\widetilde{\mathcal{L}}_\tau \leq \lambda^+$. Then the average degree of G_τ is $\delta_\tau \approx \frac{1}{2} \cdot \widetilde{\mathcal{L}}_\tau \left(\widetilde{\mathcal{D}}_\tau - 1\right)$. More precisely $\left|\delta_\tau - \frac{1}{2} \cdot \widetilde{\mathcal{L}}_\tau \left(\widetilde{\mathcal{D}}_\tau - 1\right)\right| = O\left(\frac{1}{\tau}\right)$.*

Proof. With some constant C that only depends on $\lambda^+, \delta^+, \mathrm{m}_0$ and n_0 we have that

$$\left|\delta_\tau - \frac{1}{2} \cdot \widetilde{\mathcal{L}}_\tau \left(\widetilde{\mathcal{D}}_\tau - 1\right)\right| = \frac{1}{2} \cdot \left| \frac{\mathrm{m}_0 + \tau\left(\widetilde{\mathcal{D}}_\tau - 1\right)}{\mathrm{n}_0 + \frac{\tau}{\widetilde{\mathcal{L}}_\tau}} - \widetilde{\mathcal{L}}_\tau \left(\widetilde{\mathcal{D}}_\tau - 1\right)\right| \tag{6}$$

$$= \frac{1}{2} \cdot \left| \frac{\widetilde{\mathcal{L}}_\tau \mathrm{m}_0 - \mathrm{n}_0 \widetilde{\mathcal{L}}_\tau^{\,2} \left(\widetilde{\mathcal{D}}_\tau - 1\right)}{\widetilde{\mathcal{L}}_\tau \mathrm{n}_0 + \tau} \right| \leq \frac{C}{\tau} \tag{7}$$

□

A graph G has *arboricity* a if its edges can be subdivided into a family $F_1, \ldots, F_a$ of forests over the same vertex set. Many families of graphs that have been traditionally studied have bounded arboricity, namely all classes that are closed under the so-called graph-minor operation, see Mader [1967]. From the definition follows that then necessarily $|E(G)| \leq a \cdot |V(G)|$, so G must be sparse to have low arboricity. The converse is generally not true: there are sparse graphs that have high arboricity, so having low arboricity is a stronger requirement than being sparse. For our graph genesis we obtain the following lemma. It shows that not only the density of the graphs is bounded in terms of parameters of the sequence, but also their arboricity.

Lemma 5. *Suppose that we have a distinctive sequence with $\mathcal{O} = 1$ and $\mathcal{S} = 0$. Denote the arboricity of the initial graph G_0 by a_0. Then for $\tau > 0$ the arboricity of G_τ is $a_\tau \leq \max\{a_{\tau-1}, \mathcal{L}_\tau \cdot (\mathcal{D}_\tau - 1)\}$.*

For a proof just observe that at each τ the newly created edges that link the new vertex are no more than $\mathcal{L}_\tau \cdot (\mathcal{D}_\tau - 1)$ and can each be assigned to a different forest $F_1, \ldots, F_{a_\tau}$ in the subdivision.

Restrictions to the Neighborhood of a Vertex. The aim of this section is to show that for a large variety of choices of the parameters the graph that is induced by the neighborhood of a vertex has interesting properties. In particular it contains a large subgraph that again can be obtained from the same generating process (but with other parameters). This 'recursive' local structure can then be used to bound the clustering coefficient from below. Notable exceptions from this structural property are the "*classical*" models of random graphs, in our terminology when $\mathcal{D} = 2$. On the other hand whenever $\widetilde{\mathcal{D}}$ is bounded away from 2 the cluster coefficient can be shown to be bounded in consequence.

Let $(co_{\ldots})$ be a distinctive sequence with $\mathcal{O} = 1$ and $\mathcal{S} = 0$. Let $ob \in Ob_\infty$ be an object and let $\left(co_{\ldots}^{[ob]}\right)$ be the subsequence of contexts that contain ob. Figure 3 shows the case for 6 from Figure 1.

τ'	-2	-1	0	1	2	3
ob_0	0			0	0	
ob_2	2			2		
ob_3		3	3			3
ob_5		5	5			
ob_6			6	6	6	6
ob_7					7	
ob_8						8

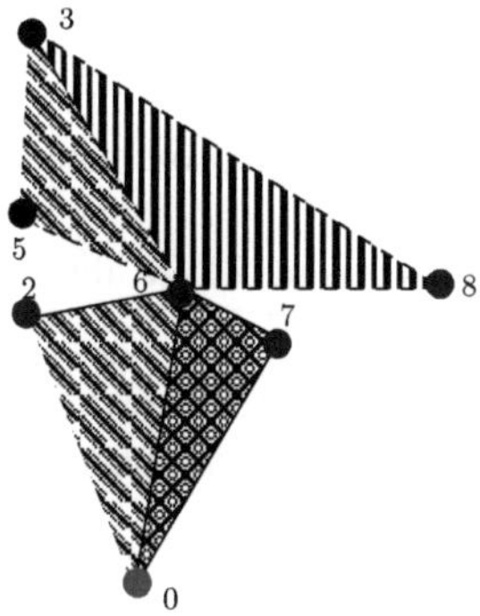

Fig. 3 The restriction $\left(co_{\dots}^{[ob_6]}\right)$ to ob_6. Only columns touching ob_6 appear and only lines corresponding to ob_6 and its neighbors

Observe that this definition only talks about the subsequence that is induced by the membership of *ob*. Other edges between the neighbors of *ob* might by induced by contexts that do not include *ob* and will thus occur sporadically in the genesis. Figure 1 illustrates such cases: when adding the two contexts for object 8 a new edge $\{7,8\}$ appears in the neighborhood of 6. Another example is edge $\{6,8\}$ that appears in the neighborhood of 0. The existence of these edges is not deducible from the contexts in which 6 (resp. 0) is involved directly. We will call such edges *locally sporadic* for the corresponding vertex. Sporadic edges may not only occur after an object has been newly introduced. In Figure 1 the edges $\{0,6\}$ and $\{6,7\}$ are sporadic for 8.

Lemma 6. *Let* $(co_{\dots})$ *be a distinctive sequence with* $\mathcal{O}=1$ *and* $\mathcal{S}=0$*. Let* $ob \in Ob_\infty$ *be an object and let* $\left(co_{\dots}^{[ob]}\right)$ *be the subsequence of contexts that contain ob. Then* $\left(co_{\dots}^{[ob]}\right)$ *is a distinctive sequence with* $\mathcal{O}=1$ *and* $\mathcal{S}=0$*, too.*

Because of this possible occurrence of locally sporadic edges the following lemma only gives a lower bound on the number of edges in the neighborhood of an object.

Lemma 7. *Let* $(co_{\dots})$ *be a distinctive sequence with* $\mathcal{O}=1$ *and* $\mathcal{S}=0$*. Let* $ob \in Ob_\infty$ *be an object, let* $\left(co_{\dots}^{[ob]}\right)$ *be the subsequence of contexts that contain ob, and let* $\mathcal{L}^{[ob]}$ *and* $\mathcal{D}^{[ob]}$ *the respective parameters of that subsequence. Then the number of edges in the neighborhood of ob at time* τ' *is at least*

$$\alpha+\beta+\tau'\left(\widetilde{\mathcal{D}^{[ob]}}_{\tau'}-2\right) \text{ with } \alpha=|E_{-1}\cap N(ob)|\,,\beta=\sum_{\tau=\tau_0}^{\tau_0+\mathcal{L}_{\tau_0}}\binom{\mathcal{D}_\tau-1}{2}. \tag{8}$$

α is the number of edges that were initially present in the neighborhood of *ob* and β is the number of edges in the neighborhood that are added before *ob* is created at time τ_0 and that would be induced by the generating contexts of *ob*.

Observe that here the time variable τ' (and thus the averaging) only accounts for those events in which *ob* is involved in the original sequence. In particular, this means that convergence to the approximate value will be much 'slower' compared to the original sequence.

Also notice that the terms in (8) are always well defined since $\mathcal{D} \geq 2$ and therefore $\widetilde{\mathcal{D}} \geq 2$, too. For contexts τ with $\mathcal{D}_\tau = 2$ the contribution to the sum in β is 0. So $\mathcal{D}_\tau = 2$ clearly is a borderline case where we have a important property change. In consequence the following theorem needs a restriction to $\mathcal{D}_\tau > 2$.

Theorem 2. *Suppose that for a distinctive sequence with* $\mathcal{O} = 1$ *and* $\mathcal{S} = 0$, *we have that* $\widetilde{\mathcal{D}}_\tau \leq \delta^+$ *and* $\widetilde{\mathcal{L}}_\tau \leq \lambda^+$ *for some constants* δ^+ *and* λ^+. *Suppose in addition that there are some integers* δ^-, λ^- *with* $2 < \delta^- \leq \mathcal{D}_\tau$ *and* $1 \leq \lambda^- \leq \mathcal{L}_\tau$ *for all* τ. *Then for* $R^- = \lambda^-(\delta^- - 2)$ *and* $R^+ = \lambda^+(\delta^+ - 1)$ *the clustering coefficient is bounded from below, namely for* τ *sufficiently large we have* $\overline{cc}(G_\tau) \geq \frac{1}{4} \cdot \frac{R^-}{R^{+2}}$.

Proof. From Lemma 4 we know that if τ is large enough (*i.e* $\tau \gg C$) at least half of the vertices have a degree of at most $2R^+$. Therefore for the clustering coefficient we get

$$\begin{aligned}
\frac{1}{n_\tau} \sum_{ob \in Ob_\tau} \frac{\mathrm{NE}_\tau(ob)}{\binom{\deg_\tau(ob)}{2}} &\geq \frac{1}{n_\tau} \sum_{\substack{ob \in Ob_\tau \\ \deg_\tau(ob) \leq 2R^+}} \frac{\mathrm{NE}_\tau(ob)}{\binom{2R^+}{2}} \\
&\geq \frac{1}{n_\tau} \sum_{\substack{ob \in Ob_\tau \\ \deg_\tau(ob) \leq 2R^+}} \frac{\lambda^-(\delta^- - 2)}{\binom{2R^+}{2}} \\
&\geq \frac{\lambda^-(\delta^- - 2)}{4R^{+2}}
\end{aligned} \tag{9}$$

And the claim follows. □

Notice that (9) only uses part of (8).

Consider the graph from Figure 1, again. Table 1 shows the statistics for the individual vertices. It turns out that the average clustering coefficient of this graph is quite high, namely 0.7. For the example graph we have $\delta^- = \delta^+ = 3$, $\lambda^- = 1$ and $\lambda^+ = 1.7$. The bound of Theorem 2 evaluates to 0.0225, which is far from the real value of 0.7.

Table 1 degree statistics of the graph in Figure 1

	$\deg_4$	NE_4	$\binom{\deg_4}{2}$	cc_4
0	5	5	10	0.5
1	2	1	1	1.0
2	3	2	3	0.7
3	4	3	6	0.5
4	2	1	1	1.0
5	3	2	3	0.7
6	6	6	15	0.4
7	3	3	3	1.0
8	4	4	6	0.7
av	3.6	3		0.7

4 Some Experimental Results

A first implementation of the random generation processes that are described here have been undertaken. They show that our approach is feasible for a large variety of parameters. It has been used to generate graphs in the range from several thousand to

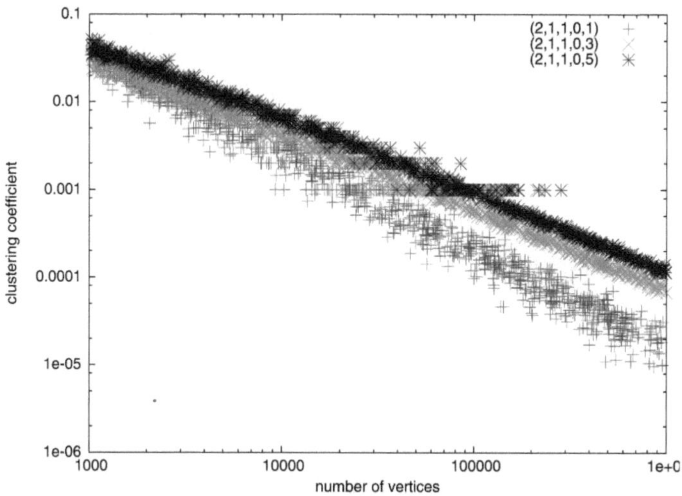

(a) $\mathcal{D} = 2$ with $\mathcal{L} = 1, 3$ and 5

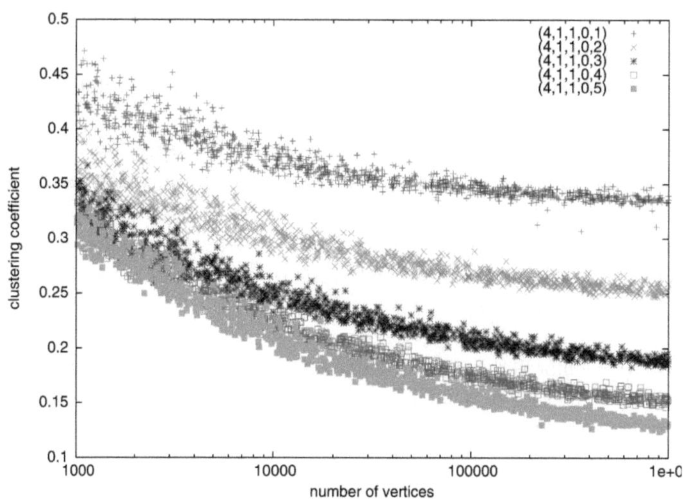

(b) $\mathcal{D} = 4$ with $\mathcal{L} = 1$ to 5

Fig. 4 Experimental results of the clustering coefficient in terms of the number of vertices (10^3 to 10^6). Each data point corresponds to a randomly generated graph G and plots the number of vertices $|V(G)|$ against the clustering coefficient $\overline{cc}(G)$. The graphs are generated such that $\log_{10}|V(G)|$ is uniformly distributed in $[3, 6)$

several million vertices. The experiments were only limited by two factors, namely the memory requirements to store the resulting graphs and the computing time that is needed for the approximation of the clustering coefficient, see Schank and Wagner [2005].

A full description of those results will be reported in a separate paper, a first preliminary report is available, see Gustedt and Schimit [2008]. Here we just like to emphasize on the threshold of $\mathcal{D} = 2$ that is apparent in Theorem 2. Figure 4 illustrates the fundamental difference in the properties of the resulting graphs for $\mathcal{D} = 2$ in Figure 4(a) (logarithmic scale in y) and $\mathcal{D} = 4$ for 4(b) (linear scale in y). For both we plot the approximation of the clustering coefficient for a set of randomly generated graphs. Whereas in the first case we have an exponential decrease of the clustering coefficient as the size of the graphs grows, the second clearly shows that we attain a non-zero limit.

5 Conclusion and Outlook

In this paper we have presented a modeling framework for the genesis of graphs. It can used for the random generation of sparse graphs for which we are able to guarantee a high local density namely a high clustering coefficient. Our approach is practical and provides quick access to large families of sample graphs that can be used for simulation and for testing.

Still, a more profound investigation of the properties of the generated graphs has to be undertaken. In particular, it will be interesting to emulate the degree and clique constraints that have been observed in different application domains. In addition, the paragon relation $\rho(\tau)$ among the contexts as it is introduced here needs more study. We have to check to what extent it models the historical bias during a graph genesis realistically.

Another direction of future research has become apparent during the first experimental studies of our approach: compared to some application graphs such as the Internet graph and social networks the graphs that we have been able to generate were quite '*small*', they had only a million vertices. To generate larger sample graphs that could be the test input for large scale tests of distributed algorithms we will need to parallelize our approach.

Acknowledgements. The author likes to thank Matthieu Latapy for interesting discussions and pointers on the subject and Pedro Schimit for a first implementation. Comments and suggestions by the anonymous reviewers have been much helpful to improve the paper for the final version.

References

Arnborg, S.: Efficient algorithms for combinatorial problems on graphs with bounded decomposability – A survey. BIT 25, 2–23 (1985)

Barabási, A.-L., Albert, R.: Emergence of scaling in random networks. Science 286, 509–512 (1999)

Boccaletti, S., Latora, V., Moreno, Y., Chavez, M., Hwang, D.-U.: Complex networks: Structure and dynamics. Physics Reports 424(4-5), 175–308 (2006), http://cogimage.dsi.cnrs.fr/publications/2006/BLMCH06

Brandes, U., Gaertler, M., Wagner, D.: Engineering graph clustering: Models and experimental evaluation. ACM Journal of Experimental Algorithms 12 (2007)

da Fontoura Costa, L., Rodrigues, F.A., Travieso, G., Villas Boas, P.R.: Characterization of complex networks: A survey of measurements. Advances In Physics 56, 167 (2007)

Dorogovtsev, S., Mendes, J.: Evolution of Networks: From Biological Nets to the Internet and WWW. Oxford University Press, Oxford (2003), http://sweet.ua.pt/~f2358/

Erdős, P., Rényi, A.: On the evolution of random graphs. Madyar Tnd. Akad. Mat. Kut. Int. Kőzl 6, 17–61 (1960)

Guillaume, J.-L., Latapy, M.: Bipartite structure of all complex networks. Information Processing Letters 90(5), 215–221 (2004)

Gustedt, J., Schimit, P.: Numerical results for generalized attachment models for the genesis of graphs. Rapport technique, INRIA (2008), http://hal.inria.fr/inria-00349461/en/ RT-0361

Latapy, M.: Grands graphes de terrain – mesure et métrologie, analyse, modélisation, algorithmique. Habilitation à diriger des recherches, Université Pierre et Marie Curie, Paris, France (2007)

Mader, W.: Homomorphieeigenschaften und mittlere Kantendichte von Graphen. Math. Ann. 174, 265–268 (1967)

Nikoloski, Z., Deo, N., Kucera, L.: Degree-correlation of scale-free graphs. In: Felsner, S. (ed.) 2005 European Conference on Combinatorics, Graph Theory and Applications (EuroComb 2005), DMTCS Proceedings. Discrete Mathematics and Theoretical Computer Science, vol. AE, pp. 239–244 (2005), http://www.dmtcs.org/dmtcs-ojs/index.php/proceedings/article/view/dmAE0

Ravasz, E., Barabási, A.-L.: Hierarchical organization in complex networks. Phys. Rev. E 67(2), 026112 (2003)

Robertson, N., Seymour, P.: Graph minors II, algorithmic aspects of tree-width. J. Algorithms 7, 309–322 (1986)

Schank, T., Wagner, D.: Approximating clustering coefficient and transitivity. J. Graph Algorithms Appl. 9(2), 265–275 (2005)

Villas Boas, P.R., Rodrigues, F.A., Travieso, G., da Fontoura Costa, L.: Chain motifs: The tails and handles of complex networks. Physical Review E 77(2), 026106 (2008), http://link.aps.org/abstract/PRE/v77/e026106

Modeling Highway Networks with Path-Geographical Transformations

Paulino Ribeiro Villas Boas, Francisco Aparecido Rodrigues,
and Luciano da Fontoura Costa

Abstract. A model of highway networks is proposed which is based on the generalization of the concept of geographical networks to incorporate several of the intermediate towns found between two main localities. This model is validated with respect to the US highway network by comparing a large number of topological measurements extracted from that structure with respective measurements obtained from ensembles of networks produced by the proposed model as well as by more traditional theoretical models of complex networks. An optimal multivariate statistical method, namely canonical analysis, is applied in order to reduce the high dimensionality of the measurements space to allow visualization as well as redundancy reduction and enhanced stochastic sampling. Maximum likelihood decision theory is then applied over the reduced measurement space in order to identify the best models. The results corroborate that the currently proposed model allows the best adherence, among all the other considered models, to the original US highway network.

1 Introduction

Complex networks have been successfully applied to model and better understand a number of real-world complex systems ranging from protein interactions to society [1]. Great part of the efforts in this dynamic research area has been applied in order to find models that can reproduce with accuracy the behavior and evolution of natural and artificial systems. Though modeling approaches often start with attempts based on existing complex networks models (e.g. Erdős-Rény [2], Barabási-Albert [3], or geographic), such models are not guaranteed to adhere to each specific problem, so that a new model may be required. Indeed, despite the large advance

Luciano da Fontoura Costa
Institute of Physics at São Carlos, University of São Paulo, PO Box 369, São Carlos, São Paulo, 13560-970 Brazil
National Institute of Science and Technology for Complex Systems
e-mail: luciano@ifsc.usp.br

R. Menezes et al. (Eds.): Complex Networks: CompleNet 2009, SCI 207, pp. 115–126.
springerlink.com © Springer-Verlag Berlin Heidelberg 2009

that has already been produced by network researchers, the structure and evolution of most complex systems are not fully understood yet [4]. In this way, the development of network models and tools to evaluate their accuracy are fundamental issues which can help to improve our understanding of complex systems.

Complex networks models have been developed for protein-protein interaction evolution [5], Internet growth [6] and human language [7], to cite just a few cases [1]. The modeling process allows forecasting the network growth as well as to perform simulations about the behavior of the system under perturbations. In addition, accurate models pave the way to perform experiments that are impossible or very expensive to be achieved otherwise, such as a simulation of a breakdown of the Internet or disease outbreaks in specific regions and conditions. The accuracy of a given model can be quantified in terms of how many structural properties of the original network it can reproduce. The larger the number of topological features that a model is capable of reproducing, the higher its accuracy. This precision can also be evaluated by considering the evolution of the original network, in the sense that the model should generate an ensemble of networks whose topological properties are similar to the original networks at any time. Figure 1 illustrates the methodology adopted in the current work for the identification, among a pre-determined set of putative models, of the model which best adheres to the original network. Evolving networks can be compared in a similar way by considering measurements obtained at each time step.

Since an accurate description of the network topology involves a large number of networks measurements [8], it is necessary to resource to multivariate statistical methods and classification approaches in order to properly evaluate the accuracy and adherence of a given model. Nevertheless, frequently it is not possible to obtain models that exactly fit a particular network structure, due to higher number of degrees of freedom, nonlinearities or very complex behavior. In such cases, the best approach is to consider several putative models and to choose the most likely one by using some statistically sound methodology. The methodology proposed in [8] allows this task to be performed with accuracy and elegance. In the current work, we propose a geographical model for highway networks, and validate it by using the just mentioned methodology.

Geographical networks are embedded in some metric space, where the vertices occupy specific positions in two or three-dimension Euclidean space and the edges represent physical connections. Examples of this important type of networks include power-grids [9], neuronal networks [10], railways connections [11], highway networks [12] and subway systems [13]. The main differences between geographical and the other complex networks, such as scale-free or random graphs, are related to the spatial constraints imposed by the distance and adjacency between vertices. Vertex degrees are limited in geographical networks, once the number of connections of a single vertex is constrained by its proximity to other vertices. The establishment of long connections is restricted by the distance-dependence cost of the edges. In addition, geographical networks do not present hierarchical structure such as the World Wide Web and actor networks [14], again as a consequence of their geographical limitations.

Several models of geographical networks involve selecting pairs of vertices, A and B, and connecting them with probability inversely proportional to the distance between them. In this work we propose a new model of geographical network where other vertices found between the extremity vertices A and B are incorporated, defining a path instead of a single edge between those two vertices. This type of connectivity is based on the way in which connections are established in several geographical systems, such as while constructing highways between cities and distributing power between several localities. A highway is usually constructed in order to connect two large cities, with the nearby smaller cities along the way being also connected in order to maximize the relationship between distance and demand. Such a model is intrinsically founded on the concept of geographic-path transformation, in which the connections are established between the cities located along the paths between pairs of cities. Another type of transformation, called path-star transformation, was recently considered in order to obtain airports networks from highway networks [15]. In that work, a sequential path in the highway network is substituted by a respective star of connections in the airport network. In the current work we propose a geographical model for highway topology and evaluate its accuracy while

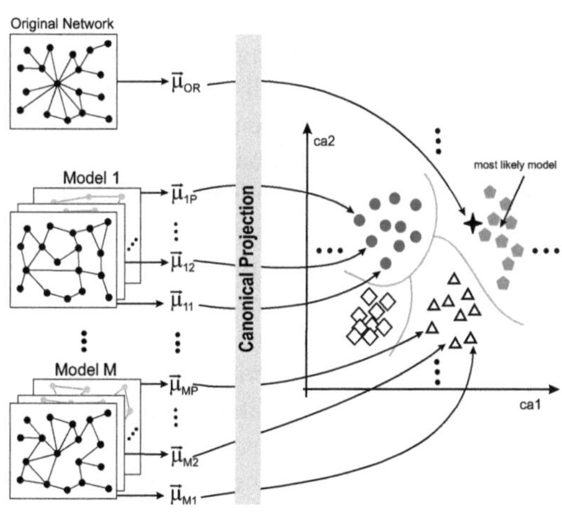

Fig. 1 The procedure for identification of the most likely model of a given original network. Several measurements are calculated for the original network ($\mu_{\mathbf{OR}}$), as well as for each of the networks in the ensembles produced by each putative model i ($\mu_{\mathbf{ij}}$, $j = 1,\ldots,P$). These measurements are projected into the 2D space in order to allow redundancy reduction, proper stochastic sample and visualization. Densities are obtained for each cluster of points in the projected space by using the Parzen windows methodology. The decision regions, whose boundary are shown by the gray lines in the projection, are obtained and the most likely model is identified as that yielding the networks which resulted closest to the original structure

reproducing the US highway network. It is shown that the proposed model yields respective networks which are the most similar to those of the original structure.

The next sections present the basic concepts about complex networks theory, as well as their measurements. Next, the classification methodology is presented with respect to the adopted concepts of canonical variable analysis and maximum likelihood decision theory. The application of the model to the US highway network is reported subsequently. The conclusion and perspective for future work are discussed in the last section.

2 Basic Concepts and Methods

2.1 Network Representation and Measurements

Complex networks are formed by a set of N vertices connected by M edges. The connections can be fully represented by the adjacency matrix A, so that each of its elements $A(i,j)$ is equal to one whenever there is a connection between i and j, with $A(i,j) = 0$ being imposed otherwise. A sequence of any adjacent edges constitutes a *walk*. On the other hand, a *path* is a walk which never re-visits any edge or vertex. In the case that a path starts and finish at the same vertex, such a closed path is called a *cycle*.

The characterization of networks can be performed in terms of several topological measurements [8]. A large set of such measurements have been considered in many works and most of them are related to the concept of connectivity and distance [16]. The immediate neighbors of a vertex i are those vertices which are connected to i. The degree of a vertex i, $k(i)$, is equal to the number of its immediate neighbors, i.e. $k(i) = \sum_j A(i,j)$. The clustering coefficient of a vertex i determines the fraction of all possible connections between the immediate neighbors of i. The average neighbor connectivity, $k_{nn}(i)$, gives the average degree of the neighbors of each vertex i in the network [17]. The global characterization of the whole network is obtained by the average of the degree, average of the clustering coefficient and the average neighbor connectivity, considering all vertices in the network.

The assortative coefficient is calculated by the degree correlation, determined from the Pearson correlation coefficient r between the degrees at both ends of the edges [18]. If $r > 0$, the network is said to be assortative (vertices with similar degrees tend to be connected); if $r < 0$, the network is disassortative (highly connected vertices tend to connect to those with few connections); for $r = 0$ there is no correlation between vertex degrees, and the network is called non-assortative.

In addition to measurements related to direct connectivity, it is also possible to characterize networks in terms of topological distances. The average shortest path length ℓ is calculated by taking into account the shortest distances between pairs of vertices. The betweenness centrality of a vertex u [19, 20] is defined as sum of the ratio between the number of shortest paths between vertices i and j that pass through vertex u and the total number of shortest paths between i and j; the sum is over all pairs i,j of distinct vertices. The average betweenness centrality $\langle B \rangle$ is calculated considering the whole set of vertices in the network and the central point dominance is defined as $c_D = \frac{1}{N-1}\sum_i(B_{\max} - B_i)$, where $B_{\max}$ represents the maximum betweenness found in the network. The central point dominance is zero for complete graphs and one for a star graphs, i.e. there is a central vertex that belongs to all paths [19].

It is also possible to extended the measurements in a more generalized, concentric fashion [21, 22]. Concentric (or hierarchical) measurements are defined by considering the successive neighborhoods around each vertex, allowing the consideration of the whole sequence of topological scales, i.e. from local neighborhood to the whole network level. In order to define such measurements, it is necessary to use the concept of ring of vertices $R_d(i)$, which is formed by those vertices which are distant d edges from the reference vertex i. The concentric degree at distance d, $k_d(i)$, is defined as the number of edges connecting the rings $R_d(i)$ and $R_{d+1}(i)$. The concentric clustering coefficient is calculated by the ratio between the number of edges in the respective d-ring, $m_d(i)$, and the maximum number of possible edges between the vertices in that ring. Note that the traditional vertex degree and clustering coefficient is a particular case of the concentric degree, corresponding to k_1 and cc_1. The divergence ratio at level d corresponds to the ratio between the number of vertices in the ring $R_d(i)$ and the hierarchical degree at distance $d-1$. In order to classify the highway network, we consider only the second concentrical level ($d = 2$). Similarly to the degree and clustering coefficient, we considered the averages of the concentric measurements as global measurements of each network.

Therefore, the adopted characterization of the original and putative networks involves the following set of nine measurements: (i) average clustering coefficient $\langle cc \rangle$, (ii) average neighborhood connectivity $\langle k_{nn} \rangle$, (iii) assortative coefficient r, (iv) average shortest path length ℓ, (v) average betweenness centrality $\langle B \rangle$, (vi) central point dominance c_D, (vii) average concentric degree of level two $\langle k_2 \rangle$, (viii) average concentric clustering coefficient of level two $\langle cc_2 \rangle$, and (ix) average divergence ratio of level two $\langle dv_2 \rangle$

3 Canonical Variable Analysis

Canonical variable analysis, which is a powerful extension of principal component analysis [23], is a technique that allows optimal projections of sets of measurements so as to obtain the reduction of the dimensionality of the original space while maximizing the separation between the known categories [24]. The canonical analysis [24, 25] is given in terms of linear combinations of the original variables. The so-called canonical variables are determined in such a way that the distances between groups (network models in our approach) are maximized relatively to the dispersion within groups (i.e. between the networks generated by each respective model). Since we adopted a set of nine measurements to characterize the networks, it is not possible to visualize the distribution of the networks in the original feature space. Another problem is that several of these measurements may be correlated, implying potential problems for the analysis. In addition, an extremely large number of simulations would be required in order to properly sample such a high dimensional measurement space so as to obtain statistically meaningful respective densities. Nevertheless, by considering the canonical variable analysis, the dimensionality reduction allows to project the networks into a two-dimensional space while reducing the correlation between such measurements and yielding more properly sampled density models. It is now possible to visualize the similarities between the networks in the sense that networks close one another in the 2D space tend to present similar features, and vice-versa.

If we consider C network models, each one identified as $C_i, i = 1, \ldots, C$, and that each network realization n is represented by its respective feature vector $\mathbf{x}_n = (x_1, x_2, \ldots, x_p)^T$, the intraclass scatter matrix [25] is defined as

$$S_{\text{intra}} = \sum_{i=1}^{C} \sum_{n \in C_i} (\mathbf{x_n} - \langle \mathbf{x} \rangle_i)(\mathbf{x_n} - \langle \mathbf{x} \rangle_i)^T, \qquad (1)$$

and the interclass scatter matrix is given as,

$$S_{\text{inter}} = \sum_{i=1}^{C} N_i (\langle \mathbf{x} \rangle_i - \langle \mathbf{x} \rangle)(\langle \mathbf{x} \rangle_i - \langle \mathbf{x} \rangle)^T, \qquad (2)$$

where $\langle \mathbf{x} \rangle_i$ corresponds to the average of the variable for the class i and $\langle \mathbf{x} \rangle$ is the general average of the given variable for all classes.

By computing the eigenvectors of the matrix $S_{\text{intra}}^{-1} S_{\text{inter}}$ and selecting those corresponding to highest absolute value eigenvalues, $\lambda_1, \ldots, \lambda_M$, it is possible to project the set of measurements into a M-dimensional space. More specifically, the canonical projection for a given network i is obtained by

$$\mathbf{X_i} = \Gamma \mathbf{x_i} \qquad (3)$$

where Γ is the transfomation matrix with the first M eigenvectors as rows.

In summary, the transformation of the original data in the measurement space into the 2D space was obtained in terms of the inner products between the measurement original feature vectors and the two eigenvectors corresponding to the highest eigenvalues.

4 Classification Approach

After obtaining the measurements and their optimal dimensionality reduction by using the canonical analysis method, the regions of classification can be determined for each complex network model by using decision theory. In the current work, we consider a supervised classification methodology which involves two stages: (i) *learning*, and (ii) *recognition*, when the trained system is used to classify new real networks. In our analysis, the training set is defined by five putative network models against which the original US highway network is compared and classified [8].

The classification regions can be obtained by the maximum likelihood approach, which can be performed by taking into account non-parametric estimation (by Parzen windows) [26] of the density probabilities of each category in the measurement space. Such an approach allows the identification of optimal decisions in situations involving uncertainty [27] in the sense of minimizing the chance of wrong classifications. Computationally, in the Parzen windows approach each point in the 2D space is considered as a Dirac's delta function. Next, these functions are convolved with a normalized Gaussian function in order to obtain the conditional probabilities for each class. More information about this approach can be found in [8, 26, 28]. After density estimation, the mass probabilities P_i, which correspond to the probability that an network belongs to class C_i, are taken into account jointly with the conditional probability densities $p(\mathbf{x_n}|C_i)$. The decision rule can be expressed as: if $P(\mathbf{x}_n|C_a)P(C_a) = \max_{b=1,m}\{P(\mathbf{x}_n|C_b)P(C_b)\}$ then select C_a,; where $\mathbf{x}_n$ is the vector that stores the network set of measurements and C_a is the class of networks associated to the model a [25, 26]. In summary, a given network n is associated to the model that results in the largest overall probability.

5 Highway Network Construction

The database of the US highway network was constructed manually considering different sources at the World Wide Web. We considered 244 cities connected by 472 highways.

6 Traditional Complex Networks Models

In order to perform the classification of the highway network, we considered four traditional network models, besides our proposed model (described in the next section).

Erdős-Rényi random graph (ER): This model generates networks with random distribution of connections. The network is constructed by connecting each pair of vertices in the network with a fixed probability p [2]. This model generates a Poisson like degree distribution.

Small-world model of Watts and Strogatz (WS): In order to construct this type of small-word network, one starts with a regular ring lattice of N vertices in which each vertex is connected to κ nearest neighbors in each direction. Each edge is then randomly rewired with probability q [29].

Barabási-Albert scale-free model (BA): This model generates networks with power law degree distribution. The network is obtained by starting with a set of m_0 vertices and, at each time step, the network grows with the addition of a new vertex with m links. The vertices which receive the new edges are chosen following a linear preferential attachment rule, i.e. the probability of the new vertex i to connect with an existing vertex j is proportional to the degree of j, $\mathscr{P}(i \to j) = k_j / \sum_u k_u$ [3].

Waxman geographical model (WG): The networks are created by considering the original position of the cities and connecting them with probability $p = e^{-\lambda d}$, where d is their geographic distance, and λ is a constant adjusted to achieve the desirable average degree [30].

7 Geographic Path Highway Network Model (GP)

Our model is based on geographical paths defined between pairs of cities. It starts with a set of isolated cities whose positions are given by the original highway network and then connects the cities according to geographical paths. For instance, in Figure 2(a), if we want to connect Midland to Denver, Phoenix, Corpus Christi, and Dallas (dashed lines), it is reasonable to connect the cities near to the respective paths so as to maximize the number of connected cities while minimizing the number and size of highways between all the cities — this is the main assumption of our model.

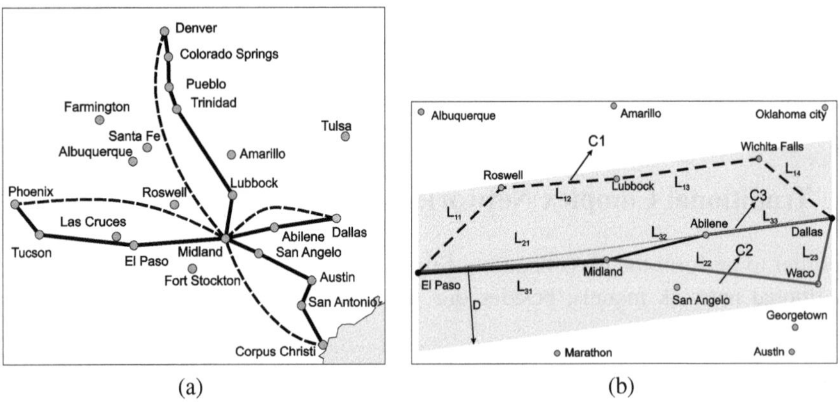

Fig. 2 (a) In our model, the connections between Midland and Phoenix, San Antonio, Denver and Dallas are established considering the cities between them, instead of constructing long range connections (indicated by dashed lines) involving only the origin and destination places. (b) Example of the methodology adopted in our model. The connection between El Paso and Dallas takes into accout only the cities inside a band of width D (indicated by the gray region). The choice of a trajectory is made considering a probability proportional to the sum of $e^{-\lambda L_e}$, where L_e is the length of its edges

This model has two stages. In the first one, all cities are considered as origin and, in the second one, origins are chosen randomly until the desired average degree is obtained. The first stage is necessary to guarantee that there is no isolated city. For both stages the destination cities are chosen according to the destination's population and the distance between departure and destination. For an origin city

S, the probability of choosing a destination city T is proportional to the product of the population of the city T and $e^{-\lambda D_{ST}}$, where λ is a parameter to control the trajectory's length and D_{ST} is the geographical distance between S and T. For larger values of λ, the chance of choosing a destination near the origin city is greater than choosing a city far away. For small values for λ, otherwise, the probability of choosing a destination far away from the origin is increased. In order to decide which towns to include along the path, a similar rule is applied. The first step is to reduce the number of possible alternative paths between the origin and destination cities, which is done by polygonally restricting the space between them, as shown in Figure 2(b). For each pair of origin and destination cities, a small rectangle of sides D_{ST} and $2D$ is draw around the closest route between origin and destination, and only the cities inside this rectangle are considered, where D is another parameter of the model used to exclude cities which are further away from the shortest route. The next step is to sort the cities inside the rectangle. All paths with more than α percentage of vertices inside the rectangle are considered. The probability of choosing one of these paths is proportional to sum of $e^{-\lambda L_e}$, where L_e is the length of the edges of each path. An example of this procedure is given in Figure 2(b), where we want to find a trajectory between the cities El Paso and Dallas. After restricting the solution space to the gray region, the cities Albuquerque, Amarillo, Oklahoma city, Marathon, Austin, and Georgetown are excluded from our possible paths, reducing the number of trajectories which can be constructed. With the remaining cities, we create possible trajectories (indicated by the dashed, black, and dark gray lines). In Figure 2(b), we can choose, for example, the trajectory indicated by the black line (smallest length), hence the obtained path is defined by the sequence: El Passo $\rightarrow$ Midland $\rightarrow$ Abilene $\rightarrow$ Dallas.

8 Results and Discussion

In order to find the best model to the highway structure, we considered the Erdős-Rényi random graph (ER), the small-world model of Watts and Strogatz (WS), the Barabási-Albert scale-free model (BA), the Waxman geographical model (WG) and our proposed geographic path model (GP). We generate 50 realizations of each

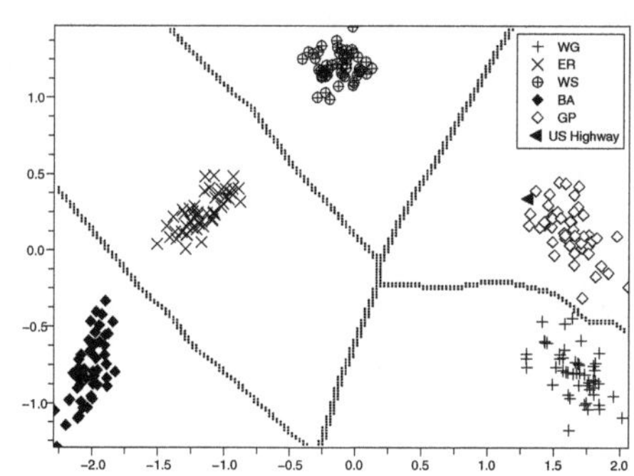

Fig. 3 Classification of the US highway network (indicated by ◀) into the considered models. Each point corresponds to one network realization and the dot lines indicate the separations between the regions of each model. See text for details

(a) (b) (c)

Fig. 4 The original US highway network (a), and examples of realization of the Waxman geographical model (b) and path geographical model (c)

model, with the same number of vertices and edges as the original highway network. The parameters used in the GP model were $\lambda = 40$ (the probability to choose destination cities near origin is much larger than those far away), $D = 0.2D_{ST}$ for the origin S and destination T of every trajectory, and $\alpha = 0.6$ (which means that 60% or more of vertices inside the rectangle of sides 2D and D_{ST} were considered).

Then, we calculated the nine measurements described in Section 2, so that each network is represented by a set of nine measurements in the original feature space. Next, we performed the canonical variable analysis and projected the whole set of networks into the 2D space (see Figure 3). It is interesting to note that the networks projections reflect the similarities between the models (e.g. the models WG and GP). Finally, by applying maximum likelihood decision theory, we determined the regions of separation, completing the training phase of the supervised classification. Then, in order to find the best model, we calculated the same set of measurements for the original US highway networks and projected the network onto the 2D space defined above, i.e. by considering the two eigenvector associated to the highest eigenvalues calculated for the network models (as in the training step). Thus, we obtained the US highway classification. As we can see in Figure 3, our GP model best reproduces the highway network topology. Although GP and WG models are nearby in Figure 3, the GP model is more similar to the US highway network than the model WG. This can also be appreciated in Figure 4, which show the original network and an example of realization of each model.

In order to provide further insights about the relation between the real network and WG and GP models, we provide the values of the measurements in Table 1. As it can be verified, only the measurement r is not reproduced by the GP model.

Table 1 Average and standard deviation (in parenthesis) of the measurements of the real network and WG and GP models

Network	$\langle cc \rangle$	$\langle k_{nn} \rangle$	ℓ	r	c_D	$\langle B \rangle$	$\langle k_2 \rangle$	$\langle cc_2 \rangle$	$\langle dv_2 \rangle$
US Highway	0.23	4.93	5.75	0.11	0.19	0.027	0.082	0.16	0.83
GN	0.15	4.47	7.78	0.30	0.18	0.015	0.095	0.11	0.78
	(0.02)	(0.17)	(1.08)	(0.07)	(0.12)	(0.009)	(0.010)	(0.01)	(0.02)
PG	0.23	4.98	5.93	-0.02	0.24	0.030	0.079	0.14	0.84
	(0.02)	(0.12)	(0.22)	(0.04)	(0.05)	(0.002)	(0.003)	(0.01)	(0.01)

9 Concluding Remarks

The basic step in growing several models of geographical networks involves choosing two vertices, A and B, and connecting them with probability inversely proportional to the distance between them. In this paper we introduced a new geographical network model where not only the link defined by A an B is incorporated into the network, but also several of the towns found between those two vertices. In other words, a path connecting A and B is included instead of only the single edge connecting A directly to B. We argue that this growth mechanism bears similarities to the way in which highways are planned and implemented, i.e. instead of establishing long connections between very populated cities, the model also connect the smaller cities found along the path, maximizing the number of visited cities while minimizing the total extension of the highways. We also investigate the potential of this new geographical model for modeling highway networks, focusing on the US highway network. This was done by taking into account several alternative theoretical models, taking a large number of topological measurements from them as well as from the original network, and applying canonical analysis and maximum likelihood decision theory in order to identify the model that best adheres to the original network. The results corroborated the ability of the new geographical model to reproduce almost all the topological features of the original network. In addition, the model can be applied in modeling other highway systems of other countries. The comparison between the parameters used for each country can help to unveil eventual universal features in highway systems.

Acknowledgements. Luciano da F. Costa is grateful to FAPESP (05/00587-5), CNPq (301303/06-1 and 573583/2008-0) for financial support. Francisco A. Rodrigues acknowledges FAPESP sponsorship (07/50633-9), Paulino R. Villas Boas acknowledges FAPESP sponsorship (08/53721-9).

References

1. da Fontoura Costa, L., Oliveira Jr, O.N., Travieso, G., Rodrigues, F.A., Villas Boas, P.R., Antiqueira, L., Viana, M.P., da Rocha, L.E.C.: Analyzing and Modeling Real-World Phenomena with Complex Networks: A Survey of Applications (2008) arXiv:0711.3199
2. Erdős, P., Rényi, A.: On random graphs. Publicationes Mathematicae 6, 290–297 (1959)
3. Barabási, A.-L., Albert, R.: Emergence of scaling in random networks. Science 286(5439), 509–512 (1999)
4. Barabási, A.-L.: The Architecture of Complexity. Control Systems Magazine, IEEE 27(4), 33–42 (2007)
5. Vázquez, A., Flammini, A., Maritan, A., Vespignani, A.: Modeling of protein interaction networks. Complexus 1(1), 38–44 (2003)
6. Yook, S.-H., Jeong, H., Barabási, A.-L.: Modeling the Internet's large-scale topology. Proceedings of the National Academy of Sciences 99(21), 13382–13386 (2002)
7. Markošová, M.: Network model of human language. Physica A 387(2-3), 661–666 (2008)

8. da Fontoura Costa, L., Rodrigues, F.A., Travieso, G., Villas Boas, P.R.: Characterization of complex networks: A survey of measurements. Advances in Physics 56(1), 167 (2007)
9. Albert, R., Albert, I., Nakarado, G.L.: Structural vulnerability of the North American power grid. Physical Review E 69(2), 025103 (2004)
10. da Fontoura Costa, L., Sporns, O.: Correlating thalamocortical connectivity and activity. Applied Physics Letters 89, 13903 (2006)
11. Seaton, K.A., Hackett, L.M.: Stations, trains and small-world networks. Physica A 339(3-4), 635–644 (2004)
12. Gastner, M.T., Newman, M.E.J.: The spatial structure of networks. The European Physical Journal B-Condensed Matter 49(2), 247–252 (2006)
13. Latora, V., Marchiori, M.: Is the Boston subway a small-world network? Physica A 314(1-4), 109–113 (2002)
14. Ravasz, E., Barabási, A.-L.: Hierarchical organization in complex networks. Physical Review E 67(2), 26112 (2003)
15. da Fontoura Costa, L.: The Path-Star Transformation and its Effects on Complex Networks (2007) arXiv:0711.1271
16. Boccaletti, S., Latora, V., Moreno, Y., Chavez, M., Hwang, D.U.: Complex networks: Structure and dynamics. Physics Reports 424(4-5), 175–308 (2006)
17. Pastor-Satorras, R., Vázquez, A., Vespignani, A.: Dynamical and correlation properties of the internet. Physical Review Letters 87(25), 258701 (2001)
18. Newman, M.E.J.: Assortative mixing in networks. Physical Review Letters 89(20), 208701 (2002)
19. Freeman, L.C.: Centrality in social networks: Conceptual clarification. Social Networks 1, 215–239 (1979)
20. Girvan, M., Newman, M.E.J.: Community structure in social and biological networks. Proceedings of the National Academy of Sciences of the United States of America 99(12), 7821–7826 (2002)
21. da Fontoura Costa, L.: The hierarchical backbone of complex networks. Physical Review Letters 93(9), 98702 (2004)
22. da Fontoura Costa, L., da Rocha, L.E.C.: A generalized approach to complex networks. The European Physical Journal B-Condensed Matter 50(1), 237–242 (2006)
23. Johnson, R.A., Wichern, D.W.: Applied Multivariate Statistical Analysis. Prentice-Hall, Englewood Cliffs (1998)
24. Campbell, N.A., Atchley, W.R.: The geometry of canonical variate analysis. Syst. Zool 30(3), 268–280 (1981)
25. da Fontoura Costa, L., Cesar Jr., R.M.: Shape Analysis and Classification: Theory and Practice. CRC Press, Boca Raton (2001)
26. Duda, R.O., Hart, P.E., Stork, D.G.: Pattern Classification. Wiley-Interscience, Hoboken (2000)
27. Bishop, C.M.: Pattern Recognition and Machine Learning (Information Science and Statistics). Springer, New York (2006)
28. McLachlan, G.J.: Discriminant analysis and statistical pattern recognition. Wiley, New York (1992)
29. Watts, D.J., Strogatz, S.H.: Collective dynamics of small-world networks. Nature 393(6684), 440–442 (1998)
30. Waxman, B.M.: Routing of multipoint connections. IEEE Journal on Selected Areas in Communications 6(9), 1617–1622 (1988)

Simplicial Complex of Opinions on Scale-Free Networks

Slobodan Maletić and Milan Rajković

Abstract. We present a model of opinions on simplicial complexes characterized by traits. The population of agents is situated at the sites of three types of scale-free networks of different sizes which display two important characteristics of real-world social networks: clustering and modularity. A new method of opinion formation is presented using advantages of simplicial complex approach. The dynamics is twofold: the agents change their opinions and new opinions are created. Topological invariants of simplicial complexes reveal relationships among opinions, particularly those which survive.

1 Introduction

In recent years the attempt to understand social dynamics from opinion and cultural dynamics to language dynamics resulted in creation of variety of models [1]. Most of the opinion formation models focus on formation of social domains of agents associated with certain property, which is changed under some dynamical mechanism controlled by one or more parameters. On the other hand less attention is devoted to formation of new opinions influenced by interaction of agents. Opinions are characterized by certain number of properties and people sharing these properties have similar opinions.

Taking the main idea of the Sznajd model [2] that *information flows outwards*, and that agents can take one of $o = 1, 2, ..., O$ opinions, we can formalize the following Sznajd-type of interaction: if two agents have the same opinion they convince

Slobodan Maletić
Institute of Nuclear Sciences Vinča, 11001, Belgrade, Serbia
e-mail: `supersloba@vin.bg.ac.yu`

Milan Rajković
Institute of Nuclear Sciences Vinča, 11001, Belgrade, Serbia
e-mail: `milanr@vin.bg.ac.yu`

R. Menezes et al. (Eds.): Complex Networks: CompleNet 2009, SCI 207, pp. 127–134.
springerlink.com © Springer-Verlag Berlin Heidelberg 2009

their neighbors (some or all) to take that opinion. In contrast to the original Sznajd model [2] and some of its variants (e.g. [3]) the following changes are introduced. If two neighboring agents i and j share the same opinion $\Omega(i) = \Omega(j) = \Omega$, than agent i convinces one of its neighbors k (excluding j) to take opinion Ω, and agent j convinces one of its neighbors l (excluding i) to take opinion Ω, regardless of the initial opinions of agents k and l. This kind of interaction we call *Sznajd type interaction*. In contrast to the Sznajd discrete model in the Deffuant continuous model [4] the agents take the opinions drawn from the uniform distribution between 0 and $+1$ (or between -1 and $+1$). Taking two neighboring agents, they change their opinions if their difference is smaller than threshold d. Let agents i and j have opinions $\omega(i)$ and $\omega(j)$, respectively, and that $|\omega(i) - \omega(j)| < d$; opinions are then adjusted according to: $\omega(i) = \omega(i) + \mu \cdot (\omega(j) - \omega(i))$ and $\omega(j) = \omega(j) + \mu \cdot (\omega(i) - \omega(j))$, where μ is the convergence parameter taken between 0 and 0.5. Introducing threshold condition enables the agents to interact only when their opinions are close enough. As an extension of the original model, Deffuant et al. [5] introduced another model called the *Relative Agreement model* (RA). Unlike the original Deffuant model, the RA model takes into consideration the uncertainty in opinion dynamics, that is, "the change in opinion $\omega(j)$ of agent j under the influence of agent i is proportional to the *overlap* between both segments (the agreement), divided by the uncertainty of the influencing segment". This kind of interaction we call *Relative Agreement interaction*. Stauffer et al. [6] also introduced discretized opinion dynamics of the Deffuant model. In this model a random pair of neighboring agents can convince each other only if their opinions differ by ± 1, and thus $1/O$ corresponds to the confidence interval of the previous models. Hence, if agents have opinions which differ by one unit, they talk to each other, and randomly one of them takes the opinion of the other.

We introduce here a model which unifies main properties of the above models. The agents are situated at sites of the scale-free network, and each agent takes the opinion which is characterized by a discrete number of traits. For example, if agent i has an opinion $o_1(i)$ characterized by traits $\{1,2,3,4\}$ and its neighboring agent j has an opinion $o_2(j)$ characterized by traits $\{1,2,5,6,7\}$, they have two traits in common, $\{1,2\}$. They discuss about opinions $o_1(i)$ and $o_2(j)$, and agent i changes his opinion by changing trait 3 of opinion o_1 with trait 5 of opinion o_2. By this process, agent i changed his opinion o_1 to new opinion o_3 characterized by traits $\{1,2,5,4\}$, and opinions of these two agents are now more alike. Furthermore, if they continue their discussion, both of them can take the same opinion o_4 by merging all traits, i.e. $o_4(i) = o_4(j) = \{1,2,3,4,5,6,7\}$. The overlapping property of opinions of the RA model is thus embedded into our interaction model. The similarity with the discretized Deffuant model is in the existence of discretized number of opinions which agents take, but their number is not fixed, since by interaction, new opinions may appear.

Our mechanism of opinion dynamics has certain similarities with Axelrod model of the dissemination of culture [7], as it includes two important mechanisms of the Axelrod model, social influence and homophily. Another similar model model with certain similarities is the model of naming game [8]. In contrast to the Axelrod's model and the naming game model there is a possibility of unification of opinions into larger ones in our model.

The paper is organized as follows. Section 1.2 is devoted to the definition of the model. In Section 1.3 the results of the simulation are summarized, and Section 1.4 is devoted to the conclusion and directions for the future work.

2 Model

An advantageous way of representing opinions and their overlapping relationship is by simplicial complexes [9]. Simplices are formed by vertices, and opinions are characterized by traits which create them. Representing opinions as simplices, and traits as vertices, we can create simplicial complex of opinions. By interactions between agents, new simplices (opinions) may occur. This kind of representation has several advantages. First, it gives us insight into the overall relationship between opinions associated with agents. Second, computing measures of simplicial complexes we can extract information about opinions, such as: which are the most important opinions with respect to the traits which characterize them (vertex significance), or which opinions have strong individual character among all other opinions (eccentricity). If there are opinions which are completely disintegrated from the structure ($ecc = 1$), the agents having these opinions will not change them, unless the interaction of Sznajd type takes place. Furthermore, if at the beginning of dynamics the value of the First structure vector [9] at 0-level is larger than 1 there is large probability of appearance of opinion which is completely disintegrated from the structure, unless, again, the interaction of Sznajd type takes place. If no new vertices (traits) are added into the structure in the long time limit all agents will have the same opinion characterized by all traits, with possible existence of several opinions disintegrated from the structure.

The corresponding algorithm consists of the following steps:

(1) A simplicial complex of opinions and a complex network of agents is formed.

(2) Each agent randomly takes one of the opinions.

(3) The opinions $o(i) = a$ and $o(j) = b$ of randomly taken agent i and one of its neighbors j are compared. If a and b do not overlap, move to the beginning of step (3). If a and b overlap or if they are equal the following processes may take place:

(3a) if agents i and j have the same opinion ($o(i) = o(j) = a = b$), then one of the neighbors (randomly chosen) of each of the agents takes this opinion (Sznajd type interaction);

(3b) if agents i and j have different opinions ($o(i) = a \neq b = o(j)$) but a and b share face f_{ab} one half of the dimension of each agent's opinion, randomly select one of the traits from a and a different trait from b. Randomly choose one of the agents and associate him (her) with a new opinion c, defined by all traits as in the initial opinion of agent, except the selected one which is changed by the selected trait of the other agent's opinion (overlap property of Relative Agreement model, with the threshold equal to $\frac{1}{2}$);

(3c) if simplices of agents i and j, $o(i) = a$ and $o(j) = b$, respectively, share face f_{ab} whose dimension is larger then one half of the dimension of simplices either a or b, all vertices (i.e. traits) of two simplices unify into a new simplex of opinion d, which is added into the complex (if it already does not exist), and both agents take opinion d (additional overlap property of Relative Agreement model).

The example of the processes (3b) and (3c) is the following. Let agents i and j have opinions $o(i) = \langle 1,2,3,4 \rangle$ and $o(j) = \langle 1,2,5,6,7 \rangle$, respectively. The opinion $o(i)$ has dimension $q_{o(i)} = 3$, opinion $o(j)$ has dimension $q_{o(j)} = 4$, and they share face $\langle 1,2 \rangle$ with dimension $f_{o(i)o(j)} = 1$. Since $f_{o(i)o(j)} < q_{o(i)}/2, q_{o(j)}/2$ (i.e. $1 < 3/2, 4/2$) and we randomly choose agent i's opinion and change it to $o'(i) = \langle 1,2,5,4 \rangle$. If at some other simulation step these two agents interact, they now share $\langle 1,2,5 \rangle$ with dimension $f_{o'(i)o(j)} = 2$, and dimensions of their opinions did not change. Since $f_{o'(i)o(j)} \geq q_{o'(i)}/2, q_{o(j)}/2$ (i.e. $2 \geq 3/2, 4/2$), the process (3c) takes place, and both agents take the new opinion $o''(i) = o''(j) = \langle 1,2,3,4,5,6,7 \rangle$, which has dimension $q_{o''} = 6$.

The simulation stops if none of the initial opinions present at the beginning of the simulation exist and/or there are opinions which are completely disintegrated from the structure.

3 Results of Simulation

The simulation was done for three types of scale-free networks of different sizes. The first one is Barabási-Albert type of scale-free network constructed by the algorithm introduced in [10]. The other two networks are constructed by the algorithm introduced in [11]. Model parameters used are: M - mean number of links per node, P_0 - probability of module formation, α - rewiring parameter which provide higher (smaller) clustering coefficient. Changing parameters M, P_0, and α we constructed two characteristic types of networks. For values $M = 5$, $P_0 = 0.007$, and $\alpha = 0.6$ the so called clustered-modular network was constructed. This network has well defined modular structure (that is, well defined modules or communities), as well as higher clustering coefficient. The second network we call nonclustered-modular network, with parameters $M = 5$, $P_0 = 0.007$, and $\alpha = 1$, has well defined modular structure, but significantly small clustering coefficient. The simulation was done for five network sizes: $N = 50$, 100, 200, 500, and 1000 nodes, and averaged over 100 samples.

For computational reasons we used only one initial simplicial complex of opinions which contains 5 opinions (simplices) and 11 traits (vertices) represented in Fig. 1. For the same initial simplicial complex and the same degree distribution the final (total) number of possible opinions Q associated to agents for three types of networks is approximately the same (Fig. 2a). Nevertheless, the final number of opinions associated with agents S is not the same (Fig. 2a). In Fig. 2b the number of different opinions which are associated to agents S is presented after each iteration (time). Some of S opinions are old (initial) opinions. For all three types of networks the emergence of the maximum number of opinions is evident at a certain

Fig. 1 Initial simplices of opinions (a) aggregated into the simplicial complex (b)

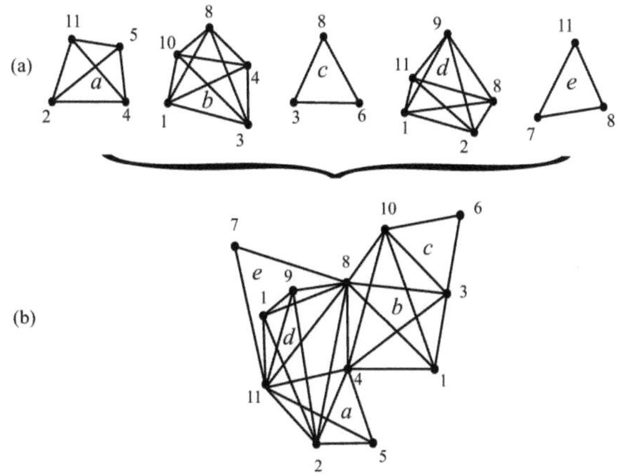

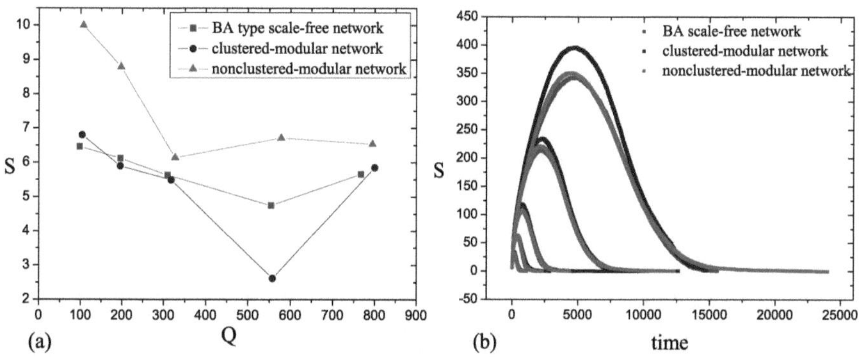

Fig. 2 (a) Dependence of the final survived opinions (S) on the final total number of formed opinions (Q) for different network sizes ($N = 50$, 100, 200, 500, and 1000). (b) Time evolution (in iteration steps) of the number of opinions associated to agents (S) for different sizes of networks ($N = 50$, 100, 200, 500, and 1000)

(characteristic) time. After that time merging of opinions into new ones becomes quicker as well as disappearance of the previously created ones. The maximum is largest for clustered-modular network. In Fig. 3a dependence of the total number of possible opinions Q (which are created by the interaction of the agents) on the size of the populations of interacting agents situated at the sites of the networks is presented. For all considered sizes of three types of networks the total number of possible opinions is approximately the same. Furthermore, very rough power-law fit yields

$$Q \sim \left(\frac{1}{N}\right)^{-\alpha} = N^{\alpha}, \tag{1}$$

where $0 < \alpha < 1$. As expected the total number of possible opinions Q is, for all three network types, an extensive quantity with a power-law dependence on the network size. This implies that connectivity has the largest impact on the total number

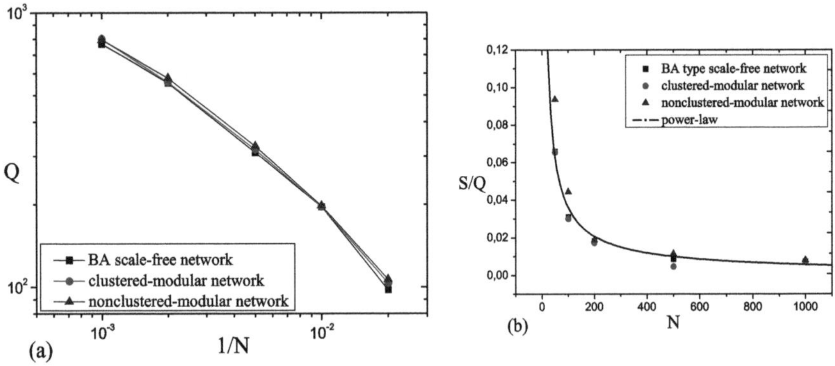

Fig. 3 (a) Dependence of the final total number of opinions (Q) on the inverse networks size ($1/N$). (b) Dependence of the scaled number of survived opinions (S/Q) on the networks size (N)

of opinions regardless of other structural properties such clustering coefficient or the tightness of the connected communities (modularity).

In Fig. 3b dependence of the ratio S/Q on the size of the network N is presented, with the power-law fit

$$\frac{S}{Q} \sim N^{-\beta}, \tag{2}$$

where $0 < \beta < 1$. Since in our case $\alpha < \beta$ ($|\alpha| \approx 0.67$ and $|\beta| \approx 0.79$), the number of survived opinions S scales as $S \sim N^{\alpha - \beta} = N^{-0.12}$ with the network size. It is important to point that this approximate result for S is obtained for relatively small network sizes, as well as for a single initial simplicial complex of opinions.

Let us consider now results for simplicial complexes of opinions for $N = 1000$. The results are obtained for three characteristic complexes taken at time t=4000 (4000-th iteration step) (a moment before the occurrence of the maximum number of opinions, see Fig. 2b). To make distinction between simplicial complexes which emerged by the dynamics on Barabási-Albert network, clustered-modular network, and nonclustered-modular network, we denote these complexes *ba-complex*, *cm-complex*, and *ncm-complex*, respectively. Hence, at $t = 4000$ *ba-complex* contains 575 opinions, *cm-complex* contains 631 opinions, and *ncm-complex* contains 558 opinions, and each of them is defined by different (or same) number of traits among 11 of them. The components of the First structure vector for all three complexes are equal to 1 and that simply means that opinions are tightly connected by sharing various number of traits. Hence, there are no isolated groups of opinions for any number of traits. Furthermore, considering the values of the Second structure vector components (Fig. 4) we can see that opinions are characterized by different numbers of traits. The eccentricity of simplices for all three complexes are equal to 0, with the exception of one simplex in each complex. These opinions are part of other opinions and hence are not integrated in the structure of opinions. On the other hand, one opinion in every complex with eccentricity different from 0 is defined by the largest number of traits, so that it has the largest dimension. Furthermore, the

Fig. 4 Components of the Second structure vector of simplicial complexes of opinions for networks of the size $N = 1000$ at time $t = 4000$ (in iteration steps)

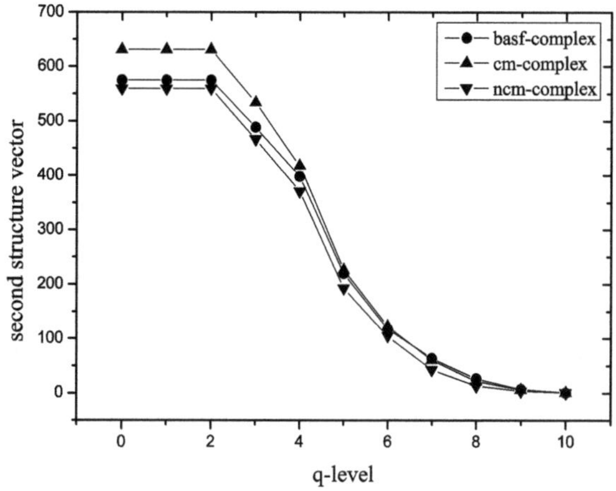

value of vertex significance for these opinion simplices is equal to 1, as they are the most significant opinions with respect to the traits which define them. The opinion simplices with $ecc = 0$ have different values of vertex significance, significantly smaller than 1. Since three opinions, one for each complex considered, are singled out among others, it is of interest to check whether they have survived at the end of simulation and, in the case they did, how many agents are associated with them. In *ba-complex* the opinion which we will call o_{473} has $ecc \neq 0$ and $vs = 1$ and is associated with 99% of agents. In *cm-complex* the opinion o_{559} has $ecc \neq 0$ and $vs = 1$ and is associated also with 99% of agents. Nevertheless, in *ncm-complex* the opinion o_{532} has $ecc \neq 0$ and $vs = 1$ and is associated with 78% of agents.

4 Conclusion

We have developed a model which represents a unification of basic concepts of two opinion dynamics models: Sznajd and RA models. The RA model was discretized with an additional property that two agents can unify their opinions into an opinion which has traits of both. We introduced the concept of creation of new opinions starting from the small set of initial opinions. Since we characterize opinions by traits and the number of traits is fixed, new opinions with possibly large number of traits, are formed through trait sharing. The interaction between agents forms new opinions which depends mainly on the size of the population of interacting agents, as well as on the total number of traits. In our model opinions correspond to simplices, and traits correspond to vertices. Topological invariants of simplicial complexes can reveal relationships between opinions in a twofold way: the relationship between similar opinions, and relationships within sets of opinions. The agents that form and exchange opinions are situated at the sites of three types of networks each of which displays power-law degree distribution (Barabási-Albert, clustered modular and non-clustered modular). The simulation results reveal that total number of possible opinions which can be associated with agents is approximately the same and depends on the network size. Nevertheless, the number of opinions which are associated with agents is significantly different for nonclustered-modular network. This discrepancy can be caused by absence of clusterization in this kind of network. It was pointed out in [6] that the final number of opinions is an extensive quantity. Naturally, the same result is valid for our simulation however the total number of opinions depends on the type of the network. On the other hand, observing time evolution (in the sense of the number of successive iterations) the maximal number of opinions which are associated with agents is largest for clustered-modular network. This discrepancy grows as the network size grows.

At the moment before reaching the maximal number of opinions associated to agents we made comparison with the final state for all three types of networks for the size $N = 1000$. First, there are no isolated groups of opinions for any number of traits. Second, opinions are characterized by different numbers of traits, as at the beginning of simulation. Third, all opinions share almost all traits with other opinions, except one opinion which is characterized by the largest number of traits. All the other opinions share all traits with this specific opinion, but none of them share all traits. Checking the existence of this particular opinion among the ones associated with the agents at the end of simulation, we found that for Barabási-Albert and clustered-modular networks almost all agents in contrast to close to $3/4$ for the nonslustered modular, take this opinion.

References

1. Castellano, C., Fortunato, S., Loreto, V.: Statistical physics of social dynamics. Rev. Mod. Phys. (in press) preprint arXiv:0710.3256
2. Sznajd-Weron, K., Sznajd, J.: Opinion evolution in closed community. Int. J. of Mod. Phys. C 11, 1157–1165 (2000)
3. Stauffer, D.: Monte Carlo simulations of Sznajd model. J. of Art. Soc. and Soc. Sim. 5(1), 4 (2001)
4. Deffuant, G., Neau, D., Amblard, F., Weisbuch, G.: Mixing beliefs among interacting agents. Adv. in Comp. Sys. 3, 87–98 (2000)
5. Deffuant, G., Amblard, F., Weisbuch, G., Faure, T.: How can extremism prevail? A study based on the relative agreement interaction model. J. of Art. Soc. and Soc. Sim. 5(4), 1 (2002)
6. Stauffer, D., Sousa, A., Schulze, C.: Discretized Opinion Dynamics of the Deffuant Model on Scale-Free Networks. J. of Art. Soc. and Soc. Sim. 7(3), 7 (2004)
7. Axelrod, R.: The Dissemination of Culture: A Model with Local Convergence and Global Polarization. J. of Confl. Res. 4(2), 203–226 (1997)
8. Dall' Asta, L., Baronchelli, A., Barrat, A., Loreto, V.: Nonequalibrium dynamics of language games on complex networks. Phys. Rev. E 74, 036105 (2006)
9. Maletić, S., Rajković, M., Vasiljević, D.: Simplicial Complexes of Networks and Their Statistical Properties. In: Bubak, M., van Albada, G.D., Dongarra, J., Sloot, P.M.A. (eds.) ICCS 2008, Part II. LNCS, vol. 5102, pp. 568–575. Springer, Heidelberg (2008)
10. Barabási, A.-L., Albert, R., Jeong, H.: Mean-field theory for scale-free random networks. Phys. A 272, 173 (1999)
11. Mitrović, M., Tadić, B.: Spectral and Dynamical Properties in Classes of Sparse Networks with Mesoscopic Inhomogeneity (preprint) arXiv:0809.4850v1

An Axiomatic Foundation for Epidemics on Complex Networks

Paul Loya and Adam R. Lucas

Abstract. We provide a rigorous axiomatic framework to study epidemiology on complex networks. Our axioms apply to the epidemic spreading on complex networks in which there are explicit correlations among the degrees of connected vertices as described in [1]. We prove a necessary and sufficient condition for our epidemic model to have a nonzero stationary solution. We believe this is the first proof of such a general result. Moreover, under appropriate conditions we show that the time independent solution is the limit of a unique time dependent solution. We also provide a rigorous definition of the epidemic threshold, $\lambda_c := 1/\lambda_1$ with λ_1 denoting the largest positive eigenvalue of an operator T given in the axioms of our model.

1 Introduction

Understanding the mechanism behind the emergence and perseverance of infected individuals on a complex network is an important problem of interest in varied disciplines including biology, physics, social sciences and mathematics [2, 3]. From the *Susceptible-Infected-Susceptible*, (abbrev. SIS), model in epidemiology a rich mathematical theory is developing which provides insight into how a disease can spread across complex networks [4, 5]. In the SIS model on a connected undirected graph, the nodes represent individuals who are in one of two states: infected (those carrying the disease) or susceptible (those who do not have the disease yet but can

Paul Loya
Department of Mathematics, Binghamton University, Vestal Parkway East, Binghamton, NY 13902, U.S.A.
e-mail: paul@math.binghamton.edu

Adam R. Lucas
Department of Mathematics and Computer Science, Saint Mary's College of California, 1928 Saint Mary's Rd., Moraga, CA 94575, U.S.A.
e-mail: arl3@stmarys-ca.edu

R. Menezes et al. (Eds.): Complex Networks: CompleNet 2009, SCI 207, pp. 135–146.
springerlink.com © Springer-Verlag Berlin Heidelberg 2009

catch it). The edges of the graph correspond to the contacts between individuals. Only susceptible individuals in contact with one or more infected individual may become infected. Infected individuals can spontaneously become susceptible again. When the spreading rate of a disease is greater than the network's epidemic threshold, a nonzero fraction of the population becomes infected in the long term. For an uncorrelated network, the epidemic threshold is the quotient of the expected degree of the network and the expected degree squared [5].

How does one define the epidemic threshold when the degree of neighboring vertices in the network are correlated? In [1], they relate the presence or absence of an epidemic threshold to the eigenvalue spectra of certain connectivity matrices of the network. One of the goals of this paper is to put this observation on a firm mathematical foundation. To do this we create an axiomatic foundation for epidemics which generalizes the uncorrelated and correlated SIS models. Our epidemic model also provides a framework for studying more general interacting particle systems.

The organization of the paper is as follows. In Section 2 we will introduce the axioms of our epidemic model. In Section 3 we show that the classical correlated and uncorrelated SIS models satisfy the axioms of our model. We point out that the dynamics of our model (under appropriate assumptions) have a unique time dependent solution which converges to a stationary one. In Section 4 we prove the main result of the paper (see Theorem 2); a necessary and sufficient condition for the existence of a nonzero stationary solution for our epidemic model.

2 Axioms of the Epidemic Model

An epidemic model shall be described in terms of three primitive notions:

$$\Omega,\ \rho,\ \text{and } T$$

where Ω is a set, $\rho : \Omega \times [0,\infty) \to [0,1]$ is a function, and $T : L^2(\Omega) \to L^2(\Omega)$ is a continuous linear map on the Hilbert space $L^2(\Omega)$ of real-valued square integrable functions on Ω. We will first describe the intended concrete interpretation of these three primitive notions before giving the axioms.

In the study of epidemics on a graph, Ω is the set of degrees of the vertices in the graph–i.e. the number of connections the vertices make to other vertices. In this context Ω is a subset of natural numbers, $\Omega \subseteq \mathbb{N}$. The set Ω is the sample space where a probability distribution $P(k), k \in \Omega$, describes the degree (i.e. number of neighbors for each vertex) distribution of the graph.

The function $\rho : \Omega \times [0,\infty) \to [0,1]$ can be interpreted as the *epidemic state* of the system. Given $k \in \Omega$ and $t \in [0,\infty)$, the number $\rho(k,t)$ represents the probability that a degree k vertex is infected at time t.

The physical meaning of the operator $T : L^2(\Omega) \to L^2(\Omega)$ needs more explanation. Recall that a Hilbert space is a complete inner product space, where the inner product is defined by

$$\langle f, g \rangle = \int_{\Omega} f(k)\, g(k)\, dP(k).$$

We are in particular interested in the subset of $L^2(\Omega)$ consisting of probability functions $\varphi : \Omega \to [0,1]$ that assign to each degree $k \in \Omega$ the probability that that a degree k vertex is infected. We call such probability functions the *states* of the system.

The operator

$$T : L^2(\Omega) \to L^2(\Omega)$$

is assumed to be an non negative integral operator in the sense that there is a measurable function

$$\tau : \Omega \times \Omega \to [0,\infty)$$

such that

$$(Tf)(k) = \int_\Omega \tau(k,k')f(k')dP(k') \quad \text{for all } f \in L^2(\Omega) \text{ and } k \in \Omega. \tag{1}$$

We shall assume that τ satisfies the generalized *detailed balance condition*

$$\tau(k,k') = \tau(k',k) \quad \text{for all } \ k,k' \in \Omega.$$

This implies, in particular, that T is *symmetric* by Fubini's theorem. In the case where Ω is a finite set, as in the case when you have a finite graph, you can think of T as a symmetric matrix with non negative entries. Let $\varphi : \Omega \to [0,1]$ be a state of the system. Then the function $T\varphi : \Omega \to [0,\infty)$ is to be interpreted as follows. Given $k \in \Omega$, the value $(T\varphi)(k) \in [0,\infty)$ is a measure of the possibility that the node k can be infected.

2.1 Axioms on the Epidemic Model

Using the three primitive notions Ω, ρ, and T intuitively explained in the previous section we now state our axioms for an epidemic model. A system (Ω,ρ,T) shall be called an *epidemic model* if the following Axioms are satisfied.

Axiom 1: Ω is the sample space of a probability space (Ω,P), where P is a probability measure on a σ-algebra of subsets of Ω.

Axiom 2: The operator $T : L^2(\Omega) \to L^2(\Omega)$ is a compact integral operator where τ in (1) satisfies the detailed balance condition.

Axiom 3: The largest positive eigenvalue, λ_1, of T has an associated eigenvector $v : \Omega \to (0,\infty)$ that is strictly positive.

Axiom 4: The *epidemic state* of the model

$$\rho : \Omega \times [0,\infty) \to [0,1]$$

satisfies the system of nonlinear differential equations

$$\boxed{\begin{cases} \partial_t\rho(k,t) = -\rho(k,t) + \lambda(1-\rho(k,t))(T\rho)(k,t), \\ \rho(k,0) = \rho_0(k), \end{cases}} \tag{2}$$

where $\rho_0 : \Omega \to [0,1]$ is the initial state of the system, assumed to be a measurable function, and where $\lambda > 0$ is called the *(effective) rate of infection*.

Axiom 5: In the case, and only in the case when, Ω is infinite, we require that for each $\lambda > 1/\lambda_1$, the inequality

$$v \leq \lambda T\left(\frac{v}{1+\varepsilon v}\right)$$

holds for all $\varepsilon > 0$ sufficiently small.

3 The Classical SIS Model

Below we will see how the classical SIS models (uncorrelated and correlated) satisfy the axioms of our epidemic model.

3.1 The Uncorrelated SIS Model

Given a connected graph (finite or infinite) let Ω denote the set of all degrees of the nodes of the graph and suppose these nodes have the degree distribution $P(k)$. Thus, (Ω, P) is a probability space with the probability measure given by

$$P(\omega) = \sum_{k\in\omega} P(k) \quad \text{for all } \ \omega \subseteq \Omega; \tag{3}$$

here we follow the standard notation of identifying the probability measure P (on the left in (3)) with its distribution $P(k)$ (on the right in (3)). The average degree of a node is given by $\langle k\rangle = \sum_i iP(i)$. Let $\rho(k,t)$ be the density of infected degree k nodes at time t. We assume that the degrees of each node are uncorrelated and hence the probability that an edge leads to an infected node is independent of the degrees of neighboring nodes. Consider the following proposition proved elsewhere [6, 7].

Proposition 1. *The probability that an edge leads to an infected node is given by*

$$\Theta = \frac{1}{\langle k\rangle}\sum_{k'} k'P(k')\rho(k',t). \tag{4}$$

Since the higher the degree of a node, the more likely the node will be infected, it follows that given k, the number $k\Theta$ is a measure of the possibility that a node of degree k can be infected.

With this proposition at hand we can now set up the classical SIS model. First, we assume that

1. The proportion of susceptible nodes become infected at a rate proportional to $(1-\rho(k,t))k\Theta$.

Note that $1-\rho(k,t)$ represents the proportion of degree k nodes not infected (the susceptible nodes) while $k\Theta$, based on the remark concerning $k\Theta$ in Proposition 1, measures the possibility that a degree k node gets infected.

We also assume that

2. The number of infected nodes decreases at a rate proportional to the number infected nodes.

The following nonlinear differential equation describes the dynamics of our system [6]:

$$\begin{cases} \partial_t \rho(k,t) = -\rho(k,t) + \lambda(1-\rho(k,t))k\Theta(t), \\ \Theta(t) = \dfrac{1}{\langle k\rangle}\displaystyle\sum_k kP(k)\rho(k,t), \\ \rho(k,0) = \rho_0(k), \end{cases} \tag{5}$$

where ρ_0 is the initial infectivity distribution.

The uncorrelated SIS model and equation (5) is a particular case of the epidemic model, (Ω,ρ,T), given in the axioms of Section 2.1. We now verify that the axioms are satisfied.

Axiom 1: Let $\Omega \subseteq \mathbb{N}$ be the set of degrees of the graph. For example, $\Omega = \{2,5,7\}$ means that every vertex has 2, 5, or 7 connections. The sigma algebra is the power set of Ω (i.e. the set of all subsets $A \subseteq \Omega$). For $A \subseteq \Omega$ we define the probability measure $P(A) = \sum_{k\in A} P(k)$ where $P(k)$ is a probability mass function on Ω.

Axiom 2: Observe that the top equation in (5) is of the form (2) if $k\Theta = T\rho$ where

$$Tf(k) = \frac{1}{\langle k\rangle}\sum_{k'} kf(k')k'P(k')$$

for all $f \in L^2(\Omega)$. In this discrete case, integrals are really summations so we can write T as

$$Tf(k) = \int_\Omega \tau(k,k')f(k')\,dP(k'),$$

where $\tau(k,k') = kk'/\langle k\rangle$. By definition, for $\tau(k,k')$ to be a measurable function, the preimage of any interval in $[0,\infty)$ must be an element of the product sigma algebra of $\Omega\times\Omega$. This is trivially satisfied here since the sigma algebra is the power set of $\Omega\times\Omega$. Moreover, it is clear that τ satisfies the detailed balance condition.

Next we show that T is a compact operator. By definition of T, given any $f \in L^2(\Omega)$ we have

$$Tf(k) = \frac{k}{\langle k\rangle}\sum_{k'} f(k')k'P(k') = \frac{1}{\langle k\rangle}\langle f,v\rangle\, v(k),$$

where $\langle f,v\rangle$ is the $L^2(\Omega)$ inner product of f and v with $v:\Omega\to(0,\infty)$ the function $v(k)=k$ for all $k\in\Omega$. Hence,

$$T = \frac{1}{\langle k \rangle} \langle \cdot, v \rangle v. \tag{6}$$

To show that T is compact we must show that if f_i is a convergent sequence of functions then $T f_i$ has a convergent subsequence. But $\frac{1}{\langle k \rangle} \langle f_i, v \rangle$ is a bounded sequence of real numbers and so by the Bolzano-Weierstrass theorem it has a convergent subsequence. Since v is fixed, it follows that $T f_i$ has a convergent subsequence.

Axiom 3: From (6) we see that the image of T is the span of v. It follows that for all $f \in L^2(\Omega)$, we have $Tf = 0$ if $f \perp v$ and if $f = v$, we have

$$Tv = \lambda_1 v\,, \quad \text{where } \lambda_1 = \frac{\langle k^2 \rangle}{\langle k \rangle}$$

with $\langle k^2 \rangle = \langle v, v \rangle = \sum_k k^2 P(k)$, the average value of k^2. It follows that T has exactly one eigenvalue, given by λ_1 and moreover, λ_1 has an associated strictly positive eigenvector $v : \Omega \to (0, \infty)$ given by $v(k) = k$ for all $k \in \Omega$.

Axiom 4: The equations in (5) are of the form (2) if $k\Theta = T\rho$.

Axiom 5: In the case when Ω is infinite, let $\lambda > 1/\lambda_1$ and observe that

$$v \le \lambda T \left(\frac{v}{1+\varepsilon v} \right) \iff v \le \frac{\lambda}{\langle k \rangle} \left\langle \frac{v}{1+\varepsilon v}, v \right\rangle v \iff 1 \le \frac{\lambda}{\langle k \rangle} \left\langle \frac{v}{1+\varepsilon v}, v \right\rangle.$$

The function $g(\varepsilon) := \frac{\lambda}{\langle k \rangle} \left\langle \frac{v}{1+\varepsilon v}, v \right\rangle$ is a continuous function of $\varepsilon \in [0, \infty)$. Since $g(0) = \frac{\lambda}{\langle k \rangle} \langle v, v \rangle = \lambda \lambda_1 > 1$, by continuity $g(\varepsilon) > 1$ for all $\varepsilon > 0$ sufficiently small. This completes the verification of Axiom 5.

3.2 *The Correlated SIS Model*

Following [1] we now allow for correlations between the degrees of neighboring vertices. We henceforth assume that Ω is a connected finite graph. Let $P(k'|k)$ be the conditional probability that a vertex of degree k' is connected to a vertex of degree k. In the spirit of Proposition 1 we have the following result proved elsewhere [1, 8].

Proposition 2

$$\Theta(k,t) = \sum_{k'} P(k'|k) \rho(k',t) \tag{7}$$

is the probability an edge originating at a degree k vertex leads to an infected vertex.

The differential equation (5) takes the following form [1]:

$$\begin{cases} \partial_t \rho(k,t) = -\rho(k,t) + \lambda(1-\rho(k,t)) k \Theta(k,t), \\ \Theta(k,t) = \sum_{k'} P(k'|k) \rho(k',t), \\ \rho(k,0) = \rho_0(k), \end{cases} \tag{8}$$

where ρ_0 is the initial infectivity distribution. We now verify that the correlated SIS system satisfies our axioms.

Axiom 1: Same as in uncorrelated case but Ω now is finite.

Axiom 2: Observe that the top equation in (8) is of the form (2) if $k\Theta = T\rho$ where

$$Tf(k) = \sum_{k'} kP(k'|k)f(k') \tag{9}$$

for all $f \in L^2(\Omega)$. Note that since Ω is finite, all functions $f : \Omega \to \mathbb{R}$ are in $L^2(\Omega)$. Hence, if N is the number of distinct degrees of the graph then $L^2(\Omega) \cong \mathbb{R}^N$ where we make the identification of a function $f : \Omega \to \mathbb{R}$ with the vector $(f(i_1), f(i_2), \ldots, f(i_N)) \in \mathbb{R}^N$ where $\Omega = \{i_1, \ldots, i_N\}$. With this identification, we see that written as a matrix we have $T = [T_{kk'}]$ where $T_{kk'} = kP(k'|k)$. Thus, we see that T is exactly the *correlation matrix* of [1]. Note that we can rewrite (9) as

$$Tf(k) = \int_{\Omega} \tau(k,k') f(k') \, dP(k'),$$

where $\tau(k,k') = kP(k'|k)/P(k')$. As in the uncorrelated case, τ is trivially a measurable function. The detailed balance condition for physical networks is [1]:

$$kP(k'|k)P(k) = k'P(k|k')P(k').$$

This equality implies that $\tau(k,k') = \tau(k',k)$. That T is a compact operator follows from that fact that $L^2(\Omega) \cong \mathbb{R}^N$ and any linear map on $\mathbb{R}^N$ is compact by the Bolzano-Weierstrass theorem.

Axiom 3: From the connectedness of our graph it follows that the operator T is irreducible and since $kP(k'|k) \geq 0$ we see from (9) that T is nonnegative, hence by the Perron-Frobenius theorem, the largest eigenvalue λ_1 of T (is simple and) has an associated strictly positive eigenvector $v : \Omega \to (0, \infty)$ (see [9] for details).

Axiom 4: The equations in (5) are of the form (2) if $k\Theta = T\rho$.

3.3 Existence and Uniqueness Theorem

The the system of nonlinear differential equations 2 does have a time dependent solution $\rho(k,t)$ which converges to the stationary one. In [8] we prove the following existence and uniqueness theorem:

Theorem 1. *With appropriate assumptions, the system of differential equations (2) has a unique solution, $\rho(k,t)$, such that*

$$\lim_{t\to\infty} \rho(k,t) = \rho_s(k),$$

where $\rho_s : \Omega \to [0,1]$ is the stationary solution to the equations (2).

As we showed in sections 3.1 and 3.2 the classical SIS models (correlated and uncorrelated) are special cases of our epidemic model. Thus, Theorem 1 applies to both of these examples. We next analyze the time-*independent* solutions.

4 The Epidemic Threshold for the General Epidemic Model

The main result of this section is the following theorem.

Theorem 2. *There exists a strictly positive stationary solution $\rho_s : \Omega \to (0,1]$ to the equations in* (2) *if and only if $\lambda > \lambda_c$ where $\lambda_c := 1/\lambda_1$ with λ_1 denoting the largest positive eigenvalue of the operator T. The number λ_c is thus referred to as the* ***epidemic threshold****.*

In particular, for the classical uncorrelated system in Section 3.1, we showed that $\lambda_1 = \langle k \rangle / \langle k^2 \rangle$, so the epidemic threshold for this system is $\lambda_c = \langle k^2 \rangle / \langle k \rangle$, which is the famous result in [6]. For the classical correlated system in Section 3.2, the epidemic threshold is given by $\lambda_c = 1/\lambda_1$ where λ_1 is the largest eigenvalue of the correlation matrix. This fact was also derived in [1] by another method.

4.1 Stationary Solutions and a Fixed Point Problem

Here, by a stationary solution, we mean a measurable function $\rho_s : \Omega \to [0,1]$ satisfying the first equation in (2) (omitting the initial condition requirement):[1]

$$0 = -\rho_s(k) + \lambda(1 - \rho_s(k))(T\rho_s)(k), \tag{10}$$

noting that $\partial_t \rho_s(k) \equiv 0$. Solving for ρ_s we find that

$$\rho_s = \frac{\lambda T \rho_s}{1 + \lambda T \rho_s}.$$

In other words, if $K = \{\text{measurable } \varphi : \Omega \to [0,1]\}$ and

$$F : K \to K$$

is the map

$$F\varphi = \frac{\lambda T \varphi}{1 + \lambda T \varphi}, \tag{11}$$

then it follows that $\rho_s : \Omega \to [0,1]$ is a stationary solution to the dynamical equation (2) if and only if $\rho_s = F\rho_s$; that is, ρ_s is a fixed point for F.

[1] Technically speaking the equality (10) only holds almost surely, that is, (10) holds except for k on a set of probability zero. However, throughout this paper, for brevity, we shall ignore sets of probability zero and write = without prefacing the equality with "almost everywhere".

Lemma 1. *For the epidemic model, the following statements are equivalent:*

1. $\rho_s : \Omega \to [0,1]$ *is a stationary solution to* (2).
2. *The function* $F : K \to K$ *has a fixed point* $\rho_s \in K$.

Of course, $\rho_s \equiv 0$ is a stationary solution, which is uninteresting, and in an epidemic theory we are really interested in *nonvanishing* stationary solutions. In the next subsection we give necessary and sufficient conditions on λ that guarantees a nonvanishing stationary solution.

4.2 Proof of Theorem 2

Assume that there is a non vanishing stationary solution ρ_s; we shall prove that $\lambda > 1/\lambda_1$. From *(2)* of Lemma 1, we know that $\rho_s = \dfrac{\lambda T\rho_s}{1+\lambda T\rho_s}$. By assumption, ρ_s is strictly positive, so $T\rho_s$ is also strictly positive, and therefore

$$\lambda T\rho_s - \rho_s = \frac{(\lambda T\rho_s)^2}{1+\lambda T\rho_s} =: g$$

defines a strictly positive function $g : \Omega \to (0,\infty)$. Hence, if $\|\ \|$ denotes the norm on $L^2(\Omega)$, then

$$\|\rho_s\|^2 = \int \rho_s(k)^2\, dP(k) < \int (\rho_s(k)+g(k))^2\, dP(k) = \|\rho_s+g\|^2.$$

Thus, $\|\rho_s\| < \|\rho_s+g\|$. On the other hand, as $\rho_s+g = \lambda T\rho_s$, we have $\|\rho_s+g\| = \lambda\|T\rho_s\|$. Since λ_1 is the largest eigenvalue of T, it follows that

$$\|T\rho_s\| \le \lambda_1\|\rho_s\|.$$

Finally, we conclude that

$$\|\rho_s\| < \lambda\lambda_1\|\rho_s\|.$$

Dividing by $\|\rho_s\| \neq 0$ we see that $1 < \lambda\lambda_1$.

Now assume that $\lambda\lambda_1 > 1$; we want to prove there is a strictly positive function $\rho_s : \Omega \to (0,1]$ such that $\rho_s = F(\rho_s)$, where $F : K \to K$ is the map found in (11).

Step 1: To find the fixed point we construct it. Let $\rho_0 := 1$ and for $i > 0$ put

$$\rho_{i+1} = F(\rho_i).$$

(Note that $1 \in K = \{\text{measurable } \varphi : \Omega \to [0,1]\}$, so $\rho_1 := F(1) \in K$, and by recursion, all the ρ_i's belong to K.) We claim that $\{\rho_i\}$ is a non increasing sequence of functions:

$$\rho_0 \ge \rho_1 \ge \rho_2 \ge \rho_3 \ge \cdots \ge 0.$$

To prove this, we first claim that F itself is non decreasing; that is, if $0 \le \varphi \le \psi$, then $F(\varphi) \le F(\psi)$. To see this, let $0 \le \varphi \le \psi$ and observe that

$$\begin{aligned}
F(\varphi) \le F(\psi) &\iff \frac{\lambda T\varphi}{1+\lambda T\varphi} \le \frac{\lambda T\psi}{1+\lambda T\psi} \\
&\iff \lambda T\varphi(1+\lambda T\psi) \le \lambda T\psi(1+\lambda T\varphi) \\
&\iff T\varphi \le T\psi \quad \text{(cancel } \lambda^2 T\varphi T\psi \text{ from both sides)} \\
&\iff \int \tau(k,k')\varphi(k')\,dP(k') \le \int \tau(k,k')\psi(k')\,dP(k'),
\end{aligned}$$

which is true since $\tau \ge 0$ (by assumption) and $\varphi \le \psi$. Now it is obvious that

$$\rho_1 := \frac{\lambda T(1)}{1+\lambda T(1)} \le 1,$$

that is, $\rho_1 \le \rho_0$. Having just proved that F is non decreasing, we see that

$$F(\rho_1) \le F(\rho_0)\,, \quad \text{which is to say,} \quad \rho_2 \le \rho_1.$$

Applying F to the inequality $\rho_2 \le \rho_1$ gives $\rho_3 \le \rho_2$. Continuing by induction we see that $\rho_{i+1} \le \rho_i$ and our proof is complete. It follows that for each $k \in \Omega$, the limit

$$\rho(k) := \lim_{i\to\infty} \rho_i(k)$$

exists.

Step 2: We claim that $\rho = F(\rho)$ and that $\rho(k) > 0$ for all $k \in \Omega$, which completes our proof. The fact that $\rho = F(\rho)$ is easy: Since $0 \le \rho_i \le 1$ for all i and the integral $\int \tau(k,k')\,1\,dP(k') = T(1)$ is finite (for almost every k), by the Lebesgue Dominated Convergence Theorem, the following interchange of limit and integral is valid:

$$\begin{aligned}
\lim_{i\to\infty} T\rho_i = \lim_{i\to\infty} \int \tau(k,k')\,\rho_i(k')\,dP(k') &= \int \tau(k,k') \lim_{i\to\infty} \rho_i(k')\,dP(k') \\
&= \int \tau(k,k')\,\rho(k')\,dP(k') \\
&= T\rho.
\end{aligned}$$

Therefore,

$$\rho := \lim_{i\to\infty} \rho_i = \lim_{i\to\infty} F(\rho_{i-1}) = \lim_{i\to\infty} \frac{\lambda T\rho_{i-1}}{1+\lambda T\rho_{i-1}} = \frac{\lambda T\rho}{1+\lambda T\rho} = F(\rho).$$

To prove that $\rho(k) > 0$ for all $k \in \Omega$, we need the following lemma.

Lemma 2. *Assume λ_1, the largest eigenvalue of T, has a strictly positive eigenvector $v : \Omega \to (0,\infty)$. For $\varepsilon > 0$ put*

$$\mathscr{A}_\varepsilon = \left\{ \varphi : \Omega \to [0,1]\,;\, \frac{\varepsilon v}{1+\varepsilon v} \le \varphi \le 1 \right\},$$

and let $F : K \to K$ be the map defined in (11)*. Then given $\lambda > 1/\lambda_1$, there is an $\varepsilon > 0$ such that*

$$F : \mathscr{A}_\varepsilon \to \mathscr{A}_\varepsilon.$$

Proof. We have to show that for $\varepsilon > 0$ sufficiently small, we have

$$\frac{\varepsilon v}{1+\varepsilon v} \le \frac{\lambda T\varphi}{1+\lambda T\varphi} \le 1 \quad \text{for all } \varphi \in \mathscr{A}_\varepsilon.$$

The second inequality is clear. By cross multiplying, the first inequality holds if and only if

$$\varepsilon v(1+\lambda T\varphi) \le \lambda T\varphi(1+\varepsilon v),$$

which holds if and only if (canceling $\varepsilon v \cdot \lambda T\varphi$ from both sides–which we can do since v is strictly positive) $\varepsilon v \le \lambda T\varphi$.

We note that T is monotone because it is given by $(Tf)(k) = \int t(k,k')f(k')dP(k')$, where $t(k,k')$ is nonnegative. So, if $f(k) \le g(k)$ it follows that $(Tf)(k) \le (Tg)(k)$. Now by monotonicity of T we have

$$\frac{\varepsilon v}{1+\varepsilon v} \le \varphi \quad \Longrightarrow \quad \lambda T\left(\frac{\varepsilon v}{1+\varepsilon v}\right) \le \lambda T\varphi.$$

Thus, to prove $\varepsilon v \le \lambda T\varphi$ holds, it suffices to prove that $\varepsilon v \le \lambda T\left(\frac{\varepsilon v}{1+\varepsilon v}\right)$ for $\varepsilon > 0$ sufficiently small; which is to say, we want to show that for $\varepsilon > 0$ sufficiently small,

$$v(k) \le \lambda T\left(\frac{v}{1+\varepsilon v}\right)(k) \quad \text{for all } k \in \Omega. \tag{12}$$

This is satisfied by Axiom 5 in the case when Ω is infinite. In the case when Ω is finite, this inequality also holds. Indeed, when $\varepsilon = 0$, the inequality (12) is $v(k) \le \lambda T(v)(k) = \lambda\lambda_1 v(k)$ which holds because $\lambda\lambda_1 > 1$. It follows by continuity that for each $k \in \Omega$ the inequality (12) holds for all $\varepsilon < \varepsilon_k$ for some $\varepsilon_k > 0$. Taking ε_0 to be the minimum of the ε_k's we see that for all $\varepsilon < \varepsilon_0$, the inequality (12) holds. This completes our proof.

Back to our proof, to see that $\rho > 0$, let v be a strictly positive eigenvector for T associated to the eigenvalue λ_1. Then by Lemma 2 we know that for $\varepsilon > 0$ sufficiently small, we have $F : \mathscr{A}_\varepsilon \to \mathscr{A}_\varepsilon$, which is to say,

$$\frac{\varepsilon v}{1+\varepsilon v} \le \varphi \le 1 \quad \Longrightarrow \quad \frac{\varepsilon v}{1+\varepsilon v} \le F(\varphi) \le 1.$$

Since $\rho_0 = 1$ in our sequence, we certainly have $\frac{\varepsilon v}{1+\varepsilon v} \le \rho_0 \le 1$. Therefore, since $\rho_1 = F(\rho_0)$, we have $\frac{\varepsilon v}{1+\varepsilon v} \le \rho_1 \le 1$. Hence, as $\rho_2 = F(\rho_1)$, again applying Lemma 2, we get $\frac{\varepsilon v}{1+\varepsilon v} \le \rho_2 \le 1$. Continuing by induction we see that $\frac{\varepsilon v}{1+\varepsilon v} \le \rho_i \le 1$ for all i. Taking $i \to \infty$ we obtain

$$0 < \frac{\varepsilon v}{1+\varepsilon v} \le \rho \le 1.$$

completing the proof.

5 Conclusions

In this paper we attempt to provide a rigorous axiomatic foundation to the study of epidemiology. The axioms of our model are given in Section 2.1. For systems that satisfy our axioms we have proved the existence of a nonzero stationary solution if and only if $\lambda > \lambda_c$ where $\lambda_c := 1/\lambda_1$ with λ_1 denoting the largest positive eigenvalue of the operator T. The number λ_c is thus referred to as the **epidemic threshold**. The classical uncorrelated and correlated SIS models are special cases of our epidemic model. Under appropriate assumptions, the nonzero stationary solution is the limit of a unique time dependant solution. Our axioms should apply to a broad range of interacting particle systems.

References

1. Boguna, M., Pastor-Satorras, R., Vespignani, A.: Epidemic spreading in complex networks with degree correlations. Lecture Notes in Physics 625, 127–147 (2003)
2. Girvan, M., Newman, M.E.J.: Community structure in social and biological networks. Proc. Natl. Acad. Sci. U.S.A. 99, 8271–8276 (2002)
3. Barabási, A.-L., Albert, R.: Emergence of scaling in random networks. Science 286(5439), 509–512 (1999)
4. Newman, M.E.J.: The spread of epidemic disease on networks. Phys. Rev. E 66, 016128 (2002)
5. Barthélemy, M., Barrat, A., Pastor-Satorras, R., Vespignani, A.: Dynamical patterns of epidemic outbreaks in complex heterogeneous networks. Journal of Theoretical Biology 235, 275–288 (2005)
6. Pastor-Satorras, R., Vespignani, A.: Epidemic dynamics and endemic states in complex networks. Phys. Rev. E 63(8), 066117 (2001)
7. Loya, P., Lucas, A.R.: On a one-parameter family of stationary solutions of the sis model (2009) (preprint)
8. Loya, P., Lucas, A.R.: Existence and uniqueness of solutions to a general epidemic model (2009) (preprint)
9. Loya, P., Lucas, A.R.: On the axiomatic foundations of epidemiology (2009) (preprint)

Analytical Approach to Bond Percolation on Clustered Networks

Sergey Melnik and James P. Gleeson

Abstract. An analytical approach to calculating bond percolation thresholds and sizes of giant connected components on random networks with non-zero clustering is presented. The networks are generated using a generalization of Trapman's [P. Trapman, Theor. Pop. Biol. **71**, 160 (2007)] model of cliques embedded in tree-like random graphs. The resulting networks have arbitrary degree distributions and tunable degree-dependent clustering. The effect of clustering on the percolation thresholds is examined and contrasted with some recent results in the literature.

1 Introduction

There has been considerable recent interest in the study of random network models, with a view to understanding the structure and dynamics of the Internet, citation networks, and other social, biological and technological networks [1, 2, 3, 4]. Random networks with a specified degree distribution P_k may be generated using the so-called *configuration model* [5]. The properties of networks generated in this manner are now well understood, with analytical results relying on the fact that such networks can be approximated very accurately by tree-like graphs [5].

However, most real-world networks are not tree-like, since the density of cycles (loops) of length three in such networks is non-zero. Because analytical results are difficult to obtain for networks containing loops, the question of how models incorporating both P_k and non-zero *degree-dependent clustering* c_k [6, 7] (taken, for example, from real-world network data) differ in structure and dynamics from corresponding randomly-wired networks (where $c_k \to 0$) remains of considerable interest.

James P. Gleeson
Department of Mathematics & Statistics, University of Limerick, Ireland
e-mail: sergey.melnik@ul.ie

R. Menezes et al. (Eds.): Complex Networks: CompleNet 2009, SCI 207, pp. 147–159.
springerlink.com © Springer-Verlag Berlin Heidelberg 2009

The bond percolation problem on networks depends strongly on the structure of the underlying graph and has several important applications in epidemiology, where the *bond occupation probability* p is related to the average transmissibility of a disease and the *giant connected component* (GCC) represents the size of an epidemic outbreak, and also in the analysis of technological networks, where the resilience of a network to the random failure of links is quantified by the size of the GCC [8].

The percolation threshold and the GCC size may be determined analytically for configuration model networks [9]. Investigations into the effects of clustering on bond percolation have also been undertaken. Newman [10] examined an example of a network in which the existence of clustering decreases the percolation threshold from its value in an unclustered network, see also [11]. Serrano and Boguñá [12, 6, 8] distinguish between strong and weak clustering with the boundary defined by $c_k = 1/(k-1)$. Their general conclusion is that weak clustering increases the percolation threshold above its unclustered value, while strong clustering decreases the threshold. While the latter conclusion is consistent with the example examined by Newman [10], it has been pointed out in the epidemiological literature [13, 14] that in clustered networks infection tends to be confined within highly connected groups, and so sufficient clustering should increase the epidemic threshold.

Trapman [15, 16] recently introduced a model of clustering in random graphs based on embedding cliques (complete subgraphs) within a tree structure. In Trapman's model networks, $c_k \propto (k-2)/k$ for all $k \geq 3$ and increases with increasing degree k, which is contrary to the typically decreasing behavior $c_k \sim k^{-1}$ for large k seen in real-world networks [17, 18].

The layout of the paper is as follows. In section 2 we generalize the Trapman algorithm for generating clustered networks. In section 3 we examine the percolation threshold for bond percolation on such clustered networks, and show that clustering may either increase or decrease the percolation threshold. Section 4 describes an analytical approach to calculating the size of the giant connected component (the epidemic size). Finally, conclusions are drawn in section 5.

2 Generating the Clustered Network

Here we describe an algorithm based on that of [15, 16] which generates random networks with arbitrary degree distributions P_k and with high clustering. The algorithm can be written in three steps, as follows:

(i) An uncorrelated random network, which we call the *super-graph*, is created using the configuration model (connecting stubs at random). The *super-graph* has a degree distribution $\widetilde{P}_k$, related to the desired distribution P_k of the final network by equation (2) below. The nodes of this super-graph are called *super-individuals*.

(ii) A fraction g_k of all k-degree super-individuals (for $k \geq 3$) is tagged as *households*, and the remaining untagged super-individuals will be referred to as *bachelors* (see Figure 1(a)).

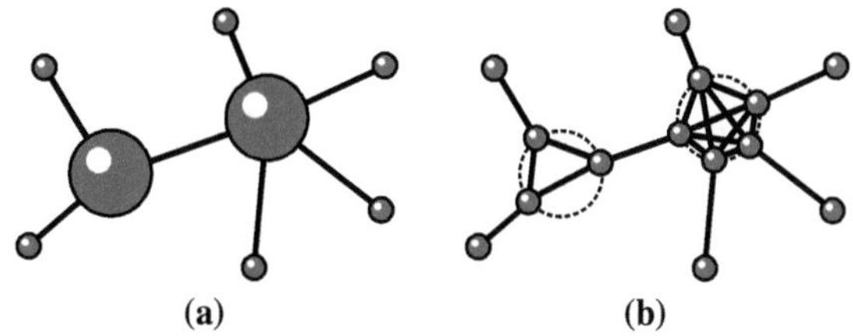

Fig. 1 (a) Graph of super-individuals which consists of two household nodes (drawn as larger nodes) and six bachelor nodes. (b) Graph of individuals which is generated from (a) by replacing households with cliques of individual nodes

(iii) Taking the tagged super-graph of step (ii) as input, we generate the *individuals graph*, in which each node represents a single individual. Each super-individual (of degree k say) which is tagged as a household is expanded into a k-clique of individual nodes as shown in Figure 1(b). Each bachelor in the super-graph becomes an individual in the individuals graph. When all super-individuals have been replaced in this way we have generated the *individuals graph* with degree distribution P_k and the algorithm concludes.

Let $\widetilde{N}$ be the total number of super-individuals in the super-graph of step (i). When $\widetilde{N}$ is sufficiently large, there are approximately $\widetilde{N}\widetilde{P}_k$ super-individuals of degree k in the network. The bachelors among these become $\widetilde{N}\widetilde{P}_k(1-g_k)$ individual nodes of degree k, while the households of degree k are expanded to $\widetilde{N}\widetilde{P}_k g_k k$ individuals grouped into k-cliques. Letting N denote the total number of individuals, we sum over all degree classes to obtain the relation $N = \widetilde{N}\sum_k \widetilde{P}_k (1 - g_k + k g_k)$.

Let f_k be the fraction of k-degree nodes in the individuals graph which are members of a k-clique. This fraction f_k is related to the fraction g_k:

$$g_k = \frac{f_k}{f_k + k - k f_k} \quad \Longleftrightarrow \quad f_k = \frac{k g_k}{1 - g_k + k g_k}. \tag{1}$$

In terms of f_k we have the following relation between the degree distributions $\widetilde{P}_k$ and P_k of the super- and individuals graphs respectively:

$$\widetilde{P}_k = \frac{P_k (1 - f_k + f_k/k)}{\sum_{k'=0}^{\infty} P_{k'} (1 - f_{k'} + f_{k'}/k')}. \tag{2}$$

The degree-dependent clustering coefficients c_k in the final, individuals graph may be calculated by noting that each k-degree individual is either a member of a single k-clique (with probability f_k) or is a member of no clique (with probability $1-f_k$). Since each node in a k-clique has clustering level $(k-2)/k$ and nodes connected using the configuration model have effectively zero clustering level in the $\widetilde{N} \to \infty$ limit (and assuming P_k has finite variance), the final average clustering for the k-degree nodes in the individuals graph may be written as

$$c_k = \frac{f_k(k-2)}{k} \qquad \text{for } k \geq 3. \tag{3}$$

Thus, given a desired degree distribution P_k and degree-dependent clustering coefficients c_k (for $k \geq 3$), the set of f_k values may be obtained from (3) with the degree distribution $\widetilde{P}_k$ and fractions g_k for the super-graph of step (i) of the algorithm following from (2) and (1) respectively. Therefore this algorithm can produce random graphs with almost any desired level of clustering (limited only by the constraint from (3) that $c_k \leq (k-2)/k$, to ensure $f_k \leq 1$). Moreover, this model gives analytically tractable results for a number of dynamical processes on networks [19]. Here we shall concentrate on the bond percolation problem. In this context it is worth noting that our algorithm, which permits k-degree nodes to be members of at most one k-clique, can be viewed as a restricted version of Newman's bipartite graph model [10]. However, unlike Newman's model, we can specify the degree distribution P_k a priori. It must be recognized that the heavily intermittent clustering due to the k-cliques gives a topological structure that may be very different to a real-world network with the same P_k and c_k; nevertheless the model can give some useful insights into the effect of clustering on GCC sizes in complex networks.

3 Percolation Threshold

3.1 Calculating p_c in Clustered Networks

The giant connected component (GCC) of an infinite graph exists if z_2, the expected number of second neighbors of a random node, exceeds z_1, the expected number of first neighbors [5]. Note both z_1 and z_2 are evaluated on the damaged graph, i.e., after a fraction $1-p$ of the links have been deleted. The lowest value of p for which $z_2/z_1 = 1$ therefore defines the percolation threshold p_c. Here we use this criterion to determine the percolation threshold (epidemic threshold) in the individuals graphs generated using the algorithm described in section 2.

Note that a GCC exists in the individuals graph if and only if the super-graph also has a GCC. It is therefore sufficient to determine a condition for the percolation transition in the super-graph, while correctly taking account of the internal k-clique structure of the super-individuals which are tagged as households.

The expected number of first neighbors in the damaged super-graph is $\widetilde{z}_1 = p\widetilde{z}$, where $\widetilde{z} = \sum k\widetilde{P}_k$ is the mean degree of the undamaged super-graph. To determine the expected number of second neighbors $\widetilde{z}_2$ in the damaged super-graph, we first choose a super-individual at random. On average, this super-individual has $\widetilde{z}_1$ first neighbors, with a given first neighbor being of degree k with probability $k\widetilde{P}_k/\widetilde{z}$. If this first neighbor is a bachelor (which occurs with probability $1-g_k$) then it connects on average to $(k-1)p$ super-individuals other than the original. If it is a household (with probability g_k) then the connections to the $(k-1)p$ further super-individuals may be thwarted by deleted internal links within the k-clique of individuals comprising the household. Thus super-individual first neighbors connect on average to $D_k(p)$ new neighbors[1], where $D_k(p)$ is a polynomial in p which may be determined exactly by methods used in [10], but whose values are bounded by

$$0 \le D_k(p) \le (k-1)p. \tag{4}$$

Combining the cases listed above, we write the expected number of second neighbors in the damaged super-graph as

$$\widetilde{z}_2 = \widetilde{z}_1 \sum_{k=1}^{\infty} \frac{k}{\widetilde{z}} \widetilde{P}_k \left((1-g_k)(k-1)p + g_k D_k(p)\right), \tag{5}$$

and so the percolation threshold p_c is the lowest value of p for which $\widetilde{z}_2/\widetilde{z}_1 = 1$, i.e. p_c satisfies the polynomial equation

$$\sum_{k=1}^{\infty} \frac{k}{\widetilde{z}} \widetilde{P}_k \left((1-g_k)(k-1)p_c + g_k D_k(p_c)\right) = 1. \tag{6}$$

Using (2) and (1) this condition may be expressed in terms of the degree-distribution P_k of the individuals graph, and the fraction f_k of k-degree individuals in cliques:

$$\sum_{k=1}^{\infty} P_k\left(k(k-1)p_c - k + f_k(k-1-k(k-1)p_c + D_k(p_c))\right) = 0. \tag{7}$$

This is a polynomial equation for the percolation threshold p_c, and its solution requires calculation of the $D_k(p)$ functions [1]. Of particular interest is the relationship between p_c and the percolation threshold in unclustered (configuration model) random networks with the same degree distribution P_k, known to be given explicitly by [9]

$$p_c^{\text{rand}} = \frac{\sum k P_k}{\sum k(k-1)P_k} = \frac{\langle k \rangle}{\langle k^2 \rangle - \langle k \rangle}. \tag{8}$$

Here the angle bracket notation denotes averaging with respect to the degree distribution P_k.

---

[1] For bond occupation probability p, the damaged k-clique may consist of a number of disconnected clusters of nodes. Letting $P(m|k)$ be the probability that a randomly chosen node in the damaged k-clique belongs to a connected cluster of m nodes (including itself), equation (7) in [10] gives $P(m|k) = \binom{k-1}{m-1}(1-p)^{m(k-m)}P(m|m)$. The probabilities $P(m|m)$ may be determined iteratively from the relation $P(k|k) = 1 - \sum_{m=1}^{k-1} P(m|k)$ with $P(1|1) = 1$. Consider an individual A in a damaged household of k individuals. We seek the number of external super-individuals which are connected to A via undamaged paths through his household—note we do not count A's own direct external link. The individual A is connected to $m-1$ other individuals in the household with probability $P(m|k)$, and each of these other individuals has a single link external to the household, which is undamaged with probability p. Thus the average number of undamaged external links from the connected cluster (and hence from A) to other super-individuals is $D_k(p) = p\sum_{m=1}^{k}(m-1)P(m|k)$.

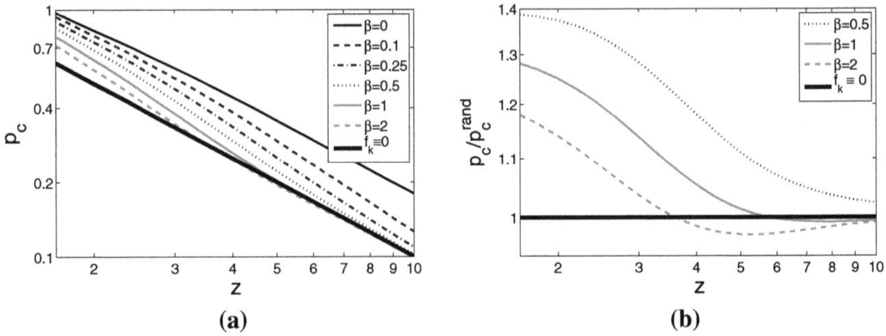

Fig. 2 (a) Percolation threshold p_c in clustered Poisson random graphs with mean degree z. The fraction $f_k = (2/(k-1))^\beta$ with β taking values indicated in the legend. The thick black curve shows the percolation threshold p_c^{rand} in the unclustered ($f_k \equiv 0$) case. (b) The ratio p_c/p_c^{rand} highlights the decrease in the percolation threshold due to clustering when β is 1 or 2

3.2 Examples

Figure 2(a) shows the percolation threshold p_c calculated from equation (7) for networks with a Poisson degree distribution $P_k = z^k e^{-z}/k!$. The log-log plots show p_c as a function of the mean degree $z = \langle k \rangle$, and for clique fractions f_k of the form

$$f_k = \left(\frac{2}{k-1}\right)^\beta \qquad \text{for } k \geq 3, \tag{9}$$

with $f_k = 0$ for $k < 3$ (since k-cliques only exist for $k \geq 3$). We show results for a set of values for β ranging from 0 (giving $f_k \equiv 1$ for all relevant k) to $\beta = 2$. Also shown (as a thick black curve) is the percolation threshold $p_c^{\text{rand}} = 1/z$ in the corresponding unclustered network. For all values of β greater than zero, we find $p_c > p_c^{\text{rand}}$ for small values of the mean degree z, but for sufficiently large z the clustered percolation point p_c becomes slightly less than the configuration model value p_c^{rand}. Figure 2(b) highlights this clustering-induced decrease of the threshold value by showing that the ratio p_c/p_c^{rand} is (slightly) less than unity for the larger z values shown.

Figure 3 shows p_c values for the truncated power-law degree distribution

$$P_k = \begin{cases} A k^{-\gamma} & , 3 \leq k \leq k_{\max} \\ 0 & , \text{otherwise} \end{cases}, \tag{10}$$

for $\gamma = 2.5$ and with the normalization constant A chosen so that $\sum P_k = 1$. The f_k dependence as in Figure 2. For convenience we have taken $P_k = 0$ for $k \leq 2$; this choice ensures the undamaged graph is relatively well-connected, and in particular that for larger values of γ a GCC exists in the unclustered network [8]. Note that here the results are presented as functions of the cutoff degree $k_{\max}$ in order to highlight

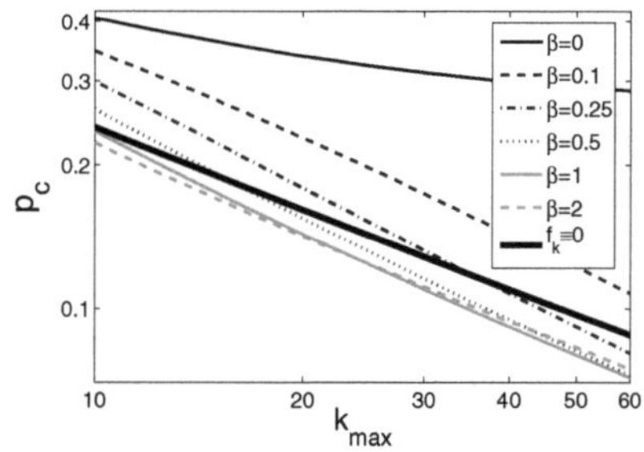

Fig. 3 Percolation threshold p_c for clustered scale-free networks with degree distribution $P_k \propto k^{-2.5}$ and cutoff degree $k_{\max}$. The fraction $f_k = (2/(k-1))^{\beta}$ with β taking values indicated in the legend

interesting behavior in the $k_{\max} \to \infty$ limit of scale-free networks. The results for the power-law degree distribution are qualitatively similar to those for the Poisson degree distribution, i.e., in all instances, except the $\beta = 0$ case of constant f_k, the clustered networks show a decrease of p_c with increasing $k_{\max}$. At large values of $k_{\max}$ we see p_c dipping below p_c^{rand} to a greater extent in Figure 3 than for the Poisson degree distribution in Figure 2.

3.3 Analytical Bounds

Some insight into these results may be gained by examining explicit bounds for p_c which may be obtained analytically from equation (7). Since $D_k(p)$ is a monotone function of p, by replacing $D_k(p)$ with its respective bounds from (4), we can solve (7) for lower and upper bounds p_- and p_+ on the value of p_c. Thus we obtain $p_- \leq p_c \leq p_+$, with

$$p_- = \frac{\langle k(1-f_k)+f_k\rangle}{\langle (k-1)(k(1-f_k)+f_k)\rangle} \quad \text{and} \quad p_+ = \frac{\langle k(1-f_k)+f_k\rangle}{\langle k(k-1)(1-f_k)\rangle}. \tag{11}$$

Note that p_- and p_+ both reduce to p_c^{rand} when $f_k \equiv 0$. We now examine the quantities $p_- - p_c^{\text{rand}}$ and $p_+ - p_c^{\text{rand}}$ for some specific forms of the clique fractions f_k. Of particular interest are cases where $p_- - p_c^{\text{rand}}$ can be shown to be positive, or where $p_+ - p_c^{\text{rand}}$ is negative. In the former case we obtain $p_c > p_- > p_c^{\text{rand}}$, and so can guarantee that the presence of such clustering increases the percolation threshold above p_c^{rand}; in the latter case we similarly guarantee that $p_c < p_c^{\text{rand}}$. After a little manipulation, we obtain the expressions

$$p_- - p_c^{\text{rand}} = \frac{\langle k\rangle \langle k(k-1)f_k\rangle - \langle k^2\rangle \langle (k-1)f_k\rangle}{\langle k(k-1)\rangle \langle (k-1)(k(1-f_k)+f_k)\rangle}, \tag{12}$$

$$p_+ - p_c^{\text{rand}} = \frac{\langle k\rangle \langle (k^2-1)f_k\rangle - \langle k^2\rangle \langle (k-1)f_k\rangle}{\langle k(k-1)\rangle \langle k(k-1)(1-f_k)\rangle}. \tag{13}$$

As the denominators are manifestly positive, the signs of these expressions are determined by the signs of their respective numerators.

3.3.1 Clustering Increases the Percolation Threshold When f_k Is Constant

We first examine $p_- - p_c^{\mathrm{rand}}$ in the case where $f_k = F$, a constant, for all $k \geq 3$. The numerator of (12) then simplifies to

$$F\left(\left\langle k^2\right\rangle - \langle k\rangle^2 + \left\langle k^2\right\rangle (P_2 - P_0) - 2\langle k\rangle P_2\right). \tag{14}$$

For the power-law degree distribution (10) we have $P_k = 0$ for $k < 3$, and so this expression reduces to $F\,\mathrm{var}(k)$ where $\mathrm{var}(k)$ is the variance $\left\langle k^2\right\rangle - \langle k\rangle^2$ of the degree distribution. Since this is positive for any $k_{\max} > 3$, we have proven that $p_c > p_c^{\mathrm{rand}}$ for constant f_k in this case. Similarly, it can be shown that (14) is positive, and hence $p_c > p_c^{\mathrm{rand}}$, for the Poisson degree distribution. These results are consistent with the $\beta = 0$ results for p_c (thin black lines) in Figures 2(a) and 3, which never dip below the p_c^{rand} values (thick black line).

3.3.2 Clustering Decreases the Percolation Threshold If $f_k = F/(k-1)$

Next, we consider the numerator of $p_+ - p_c^{\mathrm{rand}}$ for f_k of the form $F/(k-1)$ for $k \geq 3$, with F in the range $0 < F \leq 2$. The numerator of (13) then simplifies to

$$F\left(\langle k\rangle^2 + \langle k\rangle - \left\langle k^2\right\rangle - \langle k\rangle (P_0 + 2P_1 + 3P_2) + \left\langle k^2\right\rangle (P_0 + P_1 + P_2)\right). \tag{15}$$

For the power-law degree distribution (10) this further reduces to $\langle k\rangle - \mathrm{var}(k)$, and as $k_{\max} \to \infty$ this certainly becomes negative. Specifically, for the exponent $\gamma = 2.5$ used in Figure 3, this bound guarantees that p_c is less than p_c^{rand} for $k_{\max} \geq 13$. This is consistent with the curve for $\beta = 1$ in Figure 3. For the Poisson distribution, the numerator simplifies to $Fz^4e^{-z}/2$—however, as this quantity is positive we cannot draw any strong conclusions for this case.

In summary, in this section we have derived the polynomial equation (7) for the percolation threshold p_c in the presence of clustering, and solved it numerically for some examples. Analytical bounds on the value of p_c have also been derived, and for the truncated power-law degree distribution clique fractions of the form $f_k = F$ and $f_k = F/(k-1)$ have been respectively shown to guarantee that p_c is greater than, or less than, the unclustered threshold value p_c^{rand}.

4 Calculating GCC Sizes

In this section we develop an analytical approach to calculating the GCC size in the damaged individuals graph, with particular interest to the effect of clustering.

Our method is based on a general formulation for cascade sizes on random networks, described in detail in [19]. We note that a generating function approach could also be used here, similar to [10], and such a method could yield the full distribution

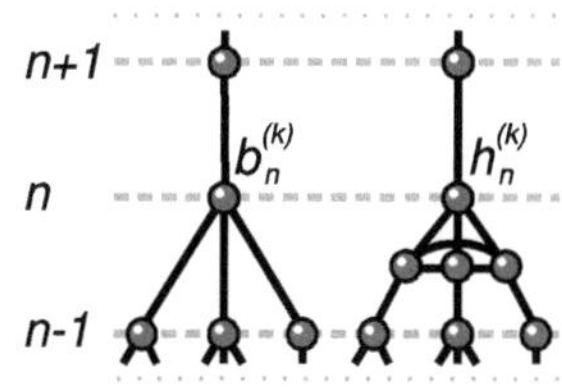

Fig. 4 Tree approximation for super-individuals graph (expanded to individuals graph). Level n is occupied by a bachelor (left) and by a top node of the expanded household (right). Other members of the same household are located at an intermediate level

of connected component sizes. However our method has the advantage of being readily generalizable to the study of other cascade-type problems on networks.

Following the approach of [19], we approximate the randomly wired super-graph as a tree structure. This tree approximation is known to be very accurate in the $\widetilde{N} \to \infty$ limit of configuration model networks. Figure 4 shows part of such a structure, with the super-individuals now expanded to show the individual nodes which constitute households. We label the levels of the tree as shown, with each super-individual at level n having a single parent at level $n+1$. Degree-k bachelors at level n therefore have $k-1$ children at level $n-1$; degree-k households at level n are considered to consist of a *top individual* (shown at level n), with the $k-1$ other individuals of the household drawn at an intermediate level. Each of these $k-1$ individuals has one child super-individual at level $n-1$.

The cascade-based approach to calculating the expected size of the GCC is as follows. Having chosen a value for the bond occupation probability p we damage the individuals graph by deleting each link between individuals with probability $1-p$. We label nodes which are part of a connected component of the graph as *active*, with the remaining nodes termed *inactive*. Define q_n as the conditional probability that a super-individual at level n is active, given that its parent at level $n+1$ is inactive. Similar conditional probabilities may be defined separately for households and for bachelors; moreover we distinguish between super-individuals of different degree k. Denote by $b_n^{(k)}$ the probability that a bachelor node of degree k at level n is active, conditional on its parent being inactive. Similarly, $h_n^{(k)}$ is the probability that the top individual node in a household of degree k is active, conditional on its parent being inactive. Since a randomly-chosen super-individual connects to a super-individual of degree k with probability $k\widetilde{P}_k/\widetilde{z}$, we have the relation

$$q_{n+1} = \sum_{k=1}^{\infty} \frac{k}{\widetilde{z}} \widetilde{P}_k \left((1-g_k) b_{n+1}^{(k)} + g_k h_{n+1}^{(k)} \right). \tag{16}$$

To determine $b_{n+1}^{(k)}$ and $h_{n+1}^{(k)}$ in terms of q_n we consider how the property of being active (i.e. being a member of a connected component) propagates from level to level. For bachelor nodes of degree k, we need to consider only their $k-1$ children at level n. Each of the children is part of a connected component with probability q_n and the link to this child is undamaged with probability p. The bachelor node becomes active if any one of the $k-1$ links to level n yield an undamaged connection to an active child, thus we have the update rule [19]

$$b_{n+1}^{(k)} = 1-(1-pq_n)^{k-1}. \tag{17}$$

For households at level $n+1$ we consider the situation of the top individual. Within the k individual nodes of the household, the top individual is part of a connected cluster of m individuals with probability $P(m|k)$ [1]. Each of the $m-1$ other individuals within the household has one edge linking to level n, and so the probability that at least one of these will become active is $1-(1-pq_n)^{m-1}$. Summing over the possible values of m, we obtain the probability of the top node of the household becoming active:

$$h_{n+1}^{(k)} = \sum_{m=1}^{k} P(m|k)\left(1-(1-pq_n)^{m-1}\right). \tag{18}$$

Combining (16), (17) and (18) enables us to write a single update equation for q_n of the form $q_{n+1} = G(q_n)$ with

$$G(q) = \sum_{k=1}^{\infty} \frac{k}{\tilde{z}}\tilde{P}_k \left[(1-g_k)\left(1-(1-pq)^{k-1}\right) + g_k \sum_{m=1}^{k} P(m|k)\left(1-(1-pq)^{m-1}\right)\right]. \tag{19}$$

Starting from an infinitesimally small value (e.g., $q_0 = 1/N$ as $N \to \infty$), this equation is iterated to yield the steady-state solution q_∞ corresponding to an infinite network. Finally, we consider the individual at the top (or root) of this infinite tree. Suppose the individual has degree k (this happens with probability P_k) and so has k children. With probability $1-f_k$ it is an individual who was a bachelor in the super-graph, and so is activated by its children with probability $1-(1-pq_\infty)^k$. Otherwise it is a member of a household of size k, and so is part of a connected cluster of m individuals within this household with probability $P(m|k)$. The whole cluster becomes active is any member of it has an undamaged link to an active child; this happens with probability $1-(1-pq_\infty)^m$. Putting together all the possibilities, we obtain an expression for S, the expected size of the giant connected component:

$$S = \sum_{k=0}^{\infty} P_k \left[(1-f_k)\left(1-(1-pq_\infty)^k\right) + f_k \sum_{m=1}^{k} P(m|k)\left(1-(1-pq_\infty)^m\right)\right], \tag{20}$$

where q_∞ is the steady-state of the iteration $q_{n+1} = G(q_n)$ defined by equation (19).

Note that a general cascade condition [19] for this system requires $dG/dq > 1$ at $q = 0$, in order that the initial iterations of the relation $q_{n+1} = G(q_n)$ allow q_n to grow finitely large. The lowest value of p for which this condition holds defines the percolation threshold p_c, and it is easy to check that this condition reduces to equation (6), which was derived using more traditional arguments in section 3.

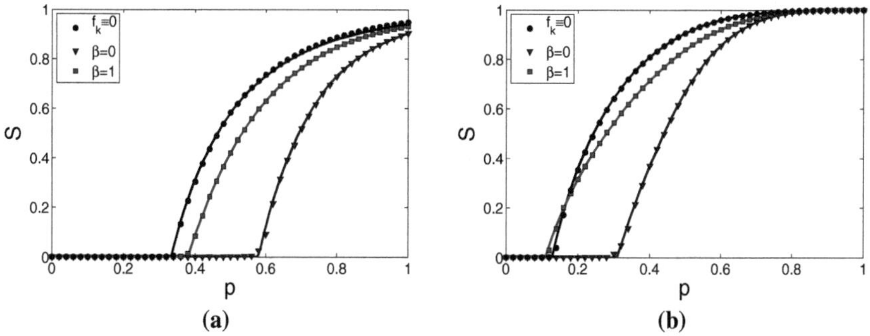

Fig. 5 Size of giant connected component S in (a) clustered Poisson random graphs with mean degree $z = 3$, and in (b) clustered graphs with truncated power-law degree distribution $P_k \propto k^{-2.5}$ for $3 \leq k \leq k_{\max}$, with $k_{\max} = 30$ here. Symbols are the results of numerical simulations on a single network with $N = 10^5$ individuals (averaged over 10 realizations of the percolation process), and curves show the analytical result from equations (16) and (20). The fraction f_k of individuals of degree k which are members of households (k-cliques) is $f_k = (2/(k-1))^\beta$ with β taking values indicated in the legend. The unclustered case ($f_k \equiv 0$) is also shown for comparison

Figures 5(a) and 5(b) show a comparison between the analytical solution (curves) and numerical computation of GCC sizes in networks generated using the algorithm of section 2 with $N = 10^5$ individuals (symbols). The degree distributions of the networks of Figure 5(a) are Poisson (as in Figure 2) with mean degree $z = 3$, while the networks for Figure 5(b) have a truncated power-law degree distribution (10) with $k_{\max} = 30$ (cf. Figure 3). Note that the values of the percolation threshold p_c predicted in Figures 2 and 3 correspond to the p values where the GCC size becomes non-zero. For the Poisson case both cases with clustering have p_c larger than the unclustered value p_c^{rand}, while the power-law case of Figure 5(b) shows that p_c may be larger or smaller than the unclustered value, depending on the form of the k-clique fraction f_k. The agreement between theory and numerical results is excellent.

5 Conclusions

We have shown that a generalization of the Trapman model [15, 16] of clustered networks has several analytically tractable features, in particular the ability to calculate the bond percolation threshold and size of the giant connected component.

The main analytical results are equation (7) for the bond percolation threshold, and the iteration scheme of section 4 (see equation (20)) for the size of the giant connected component. The percolation threshold p_c is determined by solving the polynomial equation (7), see Figures 2 and 3 for examples. We have also examined explicit upper and lower bounds for p_c (see section 3.3). Of particular interest is the relationship between p_c and the percolation threshold p_c^{rand} in a randomly-wired (unclustered) network with the same degree distribution. Our results indicate that for a given level of clustering, p_c may be greater than, or less than, p_c^{rand}, depending of the degree distribution of the network. This contrasts with the results of [8], where weakly clustered networks (with $c_k < 1/(k-1)$) have $p_c > p_c^{\text{rand}}$, while in the strongly clustered case with $c_k > 1/(k-1)$, the clustering decreases the threshold, so $p_c < p_c^{\text{rand}}$. Indeed, we show in section 3.3.1 that the Trapman model with $f_k = F$, a constant for all k, leads to clustering increasing the percolation threshold: $p_c > p_c^{\text{rand}}$, whereas the classification of this case as strongly clustered according to [8] (since $c_k = F(1-2/k)$ here) would predict the opposite conclusion.

Similarly, Figure 3 gives clear examples of cases (e.g. $\beta = 2$) where $c_k < 1/(k-1)$, but the result of $p_c < p_c^{\text{rand}}$ is the opposite to that predicted by [8] for the weakly clustered case. These contradictions to the results of [8] are not surprising when we consider that the approach of [8] is focused on clustering due to loops of length three (i.e. triangles) in the graph. Indeed, the authors of [8] carefully point out that they do not consider effects of longer loops. By contrast, the clustering within the Trapman model is more heavily localized, since a node of degree k which is a member of a triangle must also be part of a loop of length n for all n from 3 to k. Therefore we should not expect the theory of [8] to apply to the Trapman model; nevertheless it is instructive to find that model networks with the same degree distributions P_k and clustering coefficients c_k can give opposite results for this important question. Higher order information, e.g. some measure of the density of loops of length greater than three [20], is required to distinguish the two types of networks from each other.

Acknowledgements. This work was funded by Science Foundation Ireland under programmes 06/IN.1/I366, MACSI, and 05/RFP/MAT0016.

References

1. Newman, M.E.J.: SIAM Rev. 45(2), 167 (2003)
2. Dorogovtsev, S.N., Goltsev, A.V., Mendes, J.F.F.: Rev. Mod. Phys. 80(1275), 1275 (2008)
3. Boccaletti, S., Latora, V., Moreno, Y., Chavez, M., Hwang, D.U.: Phys. Rep. 424(4-5), 175 (2006)
4. Dorogovtsev, S., Mendes, J.: Evolution of Networks: From Biological Nets to the Internet and WWW. Oxford University Press, Oxford (2003)
5. Newman, M.E.J., Strogatz, S.H., Watts, D.J.: Phys. Rev. E 64(2), 026118 (2001)
6. Serrano, M.Á., Boguñá, M.: Phys. Rev. E 74(5), 056114 (2006)
7. Vázquez, A., Pastor-Satorras, R., Vespignani, A.: Phys. Rev. E 65(6), 066130 (2002)
8. Serrano, M.Á., Boguñá, M.: Phys. Rev. E 74(5), 056115 (2006)
9. Callaway, D.S., Newman, M.E.J., Strogatz, S.H., Watts, D.J.: Phys. Rev. Lett. 85(25), 5468 (2000)
10. Newman, M.E.J.: Phys. Rev. E 68(2), 026121 (2003)
11. Britton, T., Deijfen, M., Lagerås, A.N., Lindholm, M.: arXiv (2007), `http://arxiv.org/abs/0708.3939v1`
12. Serrano, M.Á., Boguñá, M.: Phys. Rev. Lett. 97(8), 088701 (2006)
13. Eames, K.T.D.: Theor. Pop. Biol. 73(1), 104 (2008)
14. Miller, J.C.: arXiv (2008), `http://arxiv.org/abs/0806.2888v1`

15. Trapman, P.: On stochastic models for the spread of infections. Ph.D. thesis, Vrije Univ. Amsterdam (2006)
16. Trapman, P.: Theor. Pop. Biol. 71(2), 160 (2007)
17. Dorogovtsev, S.N., Goltsev, A.V., Mendes, J.F.F.: Phys. Rev. E 65(6), 066122 (2002)
18. Ravasz, E., Barabási, A.L.: Phys. Rev. E 67(2), 026112 (2003)
19. Gleeson, J.P.: Phys. Rev. E 77(4), 046117 (2008)
20. Kim, H.J., Kim, J.M.: Phys. Rev. E 72(3), 036109 (2005)

Order-Wise Correlation Dynamics in Text Data

Masatoshi Funabashi, David Chavalarias, and Jean-Philippe Cointet

Abstract. We applied information geometry to treat the correlation beyond second order in network co-word analysis. We also defined the meta-level dynamics of these correlations based on all hierarchical combination of the variables, in order to encompass all possible relations. The result shows a rich variety of dynamics depending both on the order of correlation and the order of meta-level dynamics.

Keywords: Hierarchical clustering, Information geometry, co-occurrence network, political weblog, epistemic community.

1 Introduction

Network analysis is one of the prominent tool for addressing a wide range of complex systems objects. Social networks have been extensively studied since the mid twentieth century [5], lexical networks have been introduced in the 80's [2], biological systems are now commonly analyzed through network representation, etc [1, 4]. In this kind of framework, nodes represent elements, and links, eventually weighted, represent the strength of the interactions between those elements. For example, in social networks links are often assumed to represent the number of physical contacts or collaborations ; in lexical networks links could represent the number of co-occurrences of terms in a corpus ; in biological networks, links could represent

Masatoshi Funabashi
CREA (Ecole Polytechnique, France)
e-mail: Masatoshi.Funabashi@polytechnique.edu

David Chavalarias
CREA (Ecole Polytechnique, France)
e-mail: David.Chavalarias@polytechnique.edu

Jean-Philippe Cointet
CREA (Ecole Polytechnique, France)
e-mail: Jean-Philippe.Cointet@polytechnique.edu

R. Menezes et al. (Eds.): Complex Networks: CompleNet 2009, SCI 207, pp. 161–171.
springerlink.com © Springer-Verlag Berlin Heidelberg 2009

genes co-expression ; etc. It is noteworthy that in many cases, the value of links are processed over a large number of events where interactions do not take place in a one-to-one way, but rather involve potentially several elements at the same time (co-authors, co-words, etc.). This information conveyed by the patterns of interaction is lost when representing a link has an average over pairwise interactions (1). Thus, the standard way to think about links in networks only encapsulates relationships between pairs of nodes and fails to take into account an information that might be critical for further analysis. In particular, several clustering technics aim precisely at recovering communities of nodes that are supposed to interact in the same context (community of co-authors, community of words defining a scientific fields, community of genes coding for a phenotype, etc.).

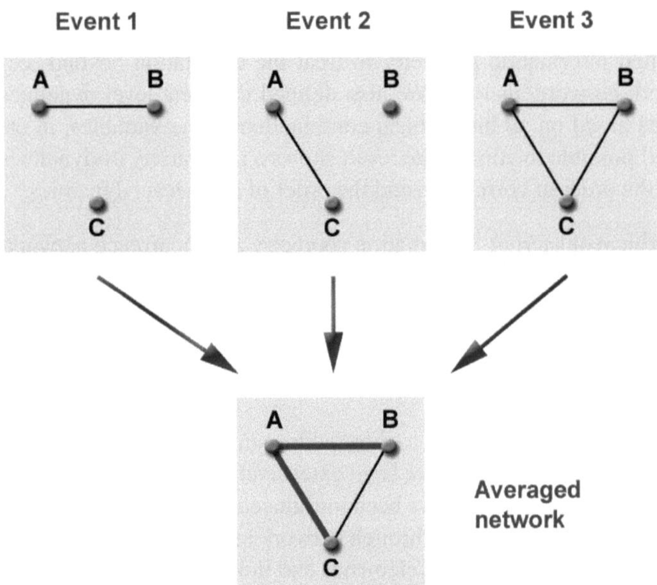

Fig. 1 In many cases, the value of links of a network are processed over events where interactions do not take place pairwise, but often involve several elements at the same time (co-authors, co-words, etc.). Thus, the information conveyed by the patterns of interaction is lost when representing a link has an average over pairwise interactions

In this paper, we propose a theoretical framework from information geometry that enables to take into account the fine grained structure of links. By distinguishing different orders of correlations between the variables associated with nodes, we consider links with the multiplicity of contexts in which they appear, and propose a method to identify the most relevant contexts. In the following, a context will be defined as a set of nodes such that their interactions are strongly correlated. Reconstructing all relevant contexts can be viewed as performing an overlapping clustering that takes into account all patterns of interaction at the event level, rather than patterns of interaction on the network structure averaged over all events. In a second section, we propose to extend this approach to higher order correlation computing meta-contexts, which is analogous to performing multi-level clustering.

2 Correlations beyond the Second Order

Although the network representation is intuitively powerful to grasp the structure of complex data , it is not appropriate to distinguish different orders of correlations between the variables associated with nodes. In real data, there exist correlations beyond the second order, which are very important to analyze collective stochastic dynamics.

Let us take for example a network with 3 nodes, each node taking binary values $\{0,1\}$ at a given time t. The dynamics of these variables as a stochastic process can be completely defined by choosing the following 7 parameters of joint distributions at time t:

$$p(1,0,0),\ p(0,1,0),\ p(0,0,1),\ p(1,1,0),\ p(1,0,1),\ p(0,1,1),\ p(1,1,1). \quad (1)$$

Note that $p(0,0,0)$ is automatically defined from the normalisation $\Sigma_{i,j,k\in\{0,1\}} p(i,j,k) = 1$.

Next, we consider the following hierarchical marginal distribution η.

$$\begin{aligned}
\eta_1 &= \sum_{i=1,\ j,k\in\{0,1\}} p(i,j,k), \quad \eta_2 = \Sigma_{j=1,\ i,k\in\{0,1\}} p(i,j,k),\\
\eta_3 &= \sum_{k=1,\ i,j\in\{0,1\}} p(i,j,k), \quad \eta_{12} = \Sigma_{i,j=1,\ k\in\{0,1\}} p(i,j,k),\\
\eta_{13} &= \sum_{i,k=1,\ j\in\{0,1\}} p(i,j,k), \quad \eta_{23} = \Sigma_{j,k=1,\ i\in\{0,1\}} p(i,j,k),\\
&\qquad\qquad \eta_{123} = \Sigma_{i,j,k=1} p(i,j,k). \quad (2)
\end{aligned}$$

η_1, η_2, and η_3 are the marginal distributions of each variable. η_{12}, η_{13}, and η_{23} represent the degrees of second-order correlations between nodes 1, 2, nodes 1,3, and nodes 2,3, respectively. Samely, η_{123} defines the degree of third-order correlation. These η can also be obtained by the Legendre transformation of natural parameters of exponential family. This fact later implies a crucial role of these coordinates.

Considering the links between nodes as a representation of interactions between nodes, it comes out straight forward that the third-order correlation included in η_{123} cannot be decided by defining the other η parameters. All the 7 parameters must be defined independently so that to specify a single model. This causes a problem to graph representation when using single kind of link between nodes, because the distinction among different orders of correlations is not considered.

The same argument also holds for the dynamics on graph with n nodes taking binary variables. The k-tuple correlation $(3 \leq k \leq n)$ can not be represented by assembling the correlations less than k-tuple ones. The higher-order correlations are independent from the lower-order ones, so that it should be measured separately for each order. (For example, even if there exists no significant second order correlation, it is possible to observe the correlation among three variables.) This class of model can be generally formulated by log-linear model, with the use of natural parameter θ.

$$\log p(x_1,x_2,\cdots,x_n) = \sum_{i=1}^{n}\theta_i x_i + \sum_{i<j}\theta_{ij}x_i x_j + \cdots + \sum_{i<j<\cdots<k}\theta_{ij\cdots k}x_i x_j\cdots x_k + \cdots + \theta_{12\cdots n}x_1 x_2\cdots x_n - \psi, \quad (3)$$

$$\psi = -\log p(0,0,\cdots,0). \quad (4)$$

For example, θ are defined as follows when $n = 3$.

$$\theta_1 = \log\frac{p(1,0,0)}{p(0,0,0)}, \qquad \theta_2 = \log\frac{p(0,1,0)}{p(0,0,0)},$$

$$\theta_3 = \log\frac{p(0,0,1)}{p(0,0,0)}, \qquad \theta_{12} = \log\frac{p(1,1,0)p(0,0,0)}{p(1,0,0)p(0,1,0)},$$

$$\theta_{13} = \log\frac{p(1,0,1)p(0,0,0)}{p(1,0,0)p(0,0,1)}, \qquad \theta_{23} = \log\frac{p(0,1,1)p(0,0,0)}{p(0,1,0)p(0,0,1)},$$

$$\theta_{123} = \log\frac{p(1,1,1)p(1,0,0)p(0,1,0)p(0,0,1)}{p(1,1,0)p(1,0,1)p(0,1,1)p(0,0,0)}. \quad (5)$$

3 Order-Wise Decomposition of Correlations

3.1 Definition of Context as Order-Wise Correlation

Although the correlation between variables can be defined up to the order of total nodes number, little is known on the correlation higher than the second order. Information geometry gives us a clear way to decompose the correlation orthogonally into different order of statistics [3]. Let p_n^n be a data distribution of n binary variables allowing up to the n-th order of correlation, and p_n^1 be the independent joint distribution with the same marginal distribution $(\eta_1,\cdots,\eta_n)$ as the data. The superscript of p represents the possible highest order of interaction, while the subscript of p is the dimension of distribution. Then the Kullback-Leibler divergence $D[p:p^1]$ between the 2 probabilities can be decomposed in the following way:

$$D[p_n^n : p_n^1] = \sum_{k=2}^{n} D_k(p_n^n), \quad (6)$$

where $D_k(p_n^n)$ represents the overall degree of interaction purely among the k variables.

The above decomposition is obtained with the use of the dual flat coordinates η and θ. The duality and the flatness of these coordinates can be defined with respect to the Fisher information matrix as metric, and dual $\alpha = \pm 1$ connections. The element $D_n(p_n^n)$, for example, is calculated by the composition of the following mixture coordinate ζ and ζ^0:

$$(\zeta_1,\cdots,\zeta_n;\zeta_{12},\cdots,\zeta_{n-1n};\cdots\cdots\cdots;\zeta_{12\cdots n-1},\cdots,\zeta_{2\cdots n};\zeta_{1\cdots n})$$
$$=(\eta_1,\cdots,\eta_n;\eta_{12},\cdots,\eta_{n-1n};\cdots\cdots\cdots;\eta_{12\cdots n-1},\cdots,\eta_{2\cdots n};\theta_{1\cdots n}), \quad (7)$$
$$(\zeta_1^0,\cdots,\zeta_n^0;\zeta_{12}^0,\cdots,\zeta_{n-1n}^0;\cdots\cdots\cdots;\zeta_{12\cdots n-1}^0,\cdots,\zeta_{2\cdots n}^0;\zeta_{1\cdots n}^0)$$
$$=(\eta_1,\cdots,\eta_n;\eta_{12},\cdots,\eta_{n-1n};\cdots\cdots\cdots;\eta_{12\cdots n-1},\cdots,\eta_{2\cdots n};0), \quad (8)$$

and

$$D_n(p_n^n)=D[\zeta:\zeta^0]. \quad (9)$$

With this example, we reduce the general form (6) to the following relation.

$$D[p_n^n:p_n^1]=D_n(p_n^n)+D[\zeta^0:p_n^1]. \quad (10)$$

We now try to isolate the arbitrary k-tuple interaction among k variables. By reducing the model to arbitrary k nodes $(2\le k\le n)$, we obtain the same relation using the reduced mixture coordinates ζ' and $\zeta^{0'}$:

$$(\zeta_1',\cdots,\zeta_k';\zeta_{12}',\cdots,\zeta_{k-1k}';\cdots\cdots\cdots;\zeta_{12\cdots k-1}',\cdots,\zeta_{2\cdots k}';\zeta_{1\cdots k}')$$
$$=(\eta_1,\cdots,\eta_k;\eta_{12},\cdots,\eta_{k-1k};\cdots\cdots\cdots;\eta_{12\cdots k-1},\cdots,\eta_{2\cdots k};\theta_{1\cdots k}), \quad (11)$$
$$(\zeta_1^{0'},\cdots,\zeta_k^{0'};\zeta_{12}^{0'},\cdots,\zeta_{k-1k}^{0'};\cdots\cdots\cdots;\zeta_{12\cdots k-1}^{0'},\cdots,\zeta_{2\cdots k}^{0'};\zeta_{1\cdots k}^{0'})$$
$$=(\eta_1,\cdots,\eta_k;\eta_{12},\cdots,\eta_{k-1k};\cdots\cdots\cdots;\eta_{12\cdots k-1},\cdots,\eta_{2\cdots k};0), \quad (12)$$
$$D_k(p_k^k)=D[\zeta':\zeta^{0'}], \quad (13)$$
$$D[p_k^k:p_k^1]=D_k(p_k^k)+D[\zeta^{0'}:p_k^1]. \quad (14)$$

Where $D_k(p_k^k)$ is the interaction purely among k variables, while $D[\zeta^{0'}:p_k^1]$ represents the rest of the interactions lower than the k-th order. Note that ζ' is not a submanifold of ζ. In ζ', we ignore completely the data outside of k nodes subgraph.

This decomposition is also compatible to chi-squared test to judge the significance of order-wise correlation. Considering only the highest order of interaction in k nodes subgraph, the following holds asymptotically with enough large sample number N.

$$\lambda=2ND_k(p_k^k)\sim\chi^2(1). \quad (15)$$

Where $\chi^2(1)$ is the chi-squared distribution with one degree of freedom.

We also consider the distinction between positive and negative correlation by changing the positive/negative sign of the statistics λ:

$$\lambda=\mathrm{sgn}(\eta_{12\cdots k}-\eta_1\eta_2\cdots\eta_k)\cdot 2ND_k(p_k^k). \quad (16)$$

Where $\mathrm{sgn}(\cdot)$ is the signum function. This means that the 5 % significance level of positive/negative correlation is judged by the λ being above/below the +/- 5 % threshold (3.841), respectively.

By changing the k, we are able to evaluate all possible sets of pure k-tuple interaction among specific set of k variables. We call this λ the significance of k-th order context among k variables. Pragmatically, the sample used to calculate λ is not necessary limited to each time step of data. In weblog data, for example, it is common to keep the history of about the past 1 week to 1 month in the same page. Therefore, the distribution at time t should also reflect these periods in the calculation by setting proportional length of the time window. This prescription also ameliorate the sensitivity of the test.

For the following analysis based on text data, we define the context as the significance of keywords co-occurence. The context of k words corresponds to the k-th order correlation of the occurrence of k words, which is evaluated by (16). Therefore, there exist $n_c = \sum_{k=1}^{n} {}_nC_k$ possible contexts in the graph of n nodes. We also define the subcontext of a context when the k-th order context of binary variables $(x_1, \cdots, x_k)$ and the k_{sub}-th order context of binary variables $(x_1, \cdots, x_{k_{sub}})$ exist, where $2 \leq k_{sub} < k$ and $\{x_1, \cdots, x_{k_{sub}}\} \subset \{x_1, \cdots, x_k\}$ are satisfied. The k_{sub}-th order context is then a subcontext of the k-th order context. This gives us a way of clustering taking the different orders of interaction into consideration. We applied this theory to distinguish different order of correlations on the log-linear model with $n = 4$ binary variables.

3.2 Definition of Meta-context

Furthermore, contexts between contexts can also be defined in the same manner. This is a kind of hierarchical clustering taking the result of the previous level. From the calculation of the n_c contexts' significance, we first judge for each order k and each period t which contexts are significant or not. Next, we obtain the occurence matrix of the contexts, that specifies with a binary variable if a context is significant (true, $= 1$) or not (false, $= 0$). This new binary data has the same format as the raw data used to calculate the contexts, except the number of variables being n_c instead of n.

It is now a question if these signifficant contexts also have correlations among each other. This information cannot be treated with context only, but requires to establish relations across contexts. It is observed in the relation between contexts. It can also be considered as the relation among n_c links of the n nodes graph. We call this co-occurrence of contexts as "meta-context." The relation between the binary variables, the contexts, and the meta-contexts is depicted in FIg. 2. Samely as the number of contexts, $n_c^{meta} = \sum_{k=1}^{n_c} {}_{n_c}C_k$ meta-contexts are possible in n nodes graph. As an expansion, performing the same operation on meta-context occurrence matrix, there exists $n_c^{meta^2} = \sum_{k=1}^{n_c^{meta}} {}_{n_c^{meta}}C_k$ meta-meta-contexts. For simplicity, we denote such expansion as meta^2-context, which is possible to generalise to the arbitrary finite levels as meta^l-context $(1 \leq l < \infty)$.

It is clear that the number of meta^l-context grows more than exponentially with l when $n \geq 4$. Though, it is natural to assume that there exists a certain limit of l, namely l_{max}, where there is no signifficant meta$^{l_{max}}$-context. Ascending the hierar-

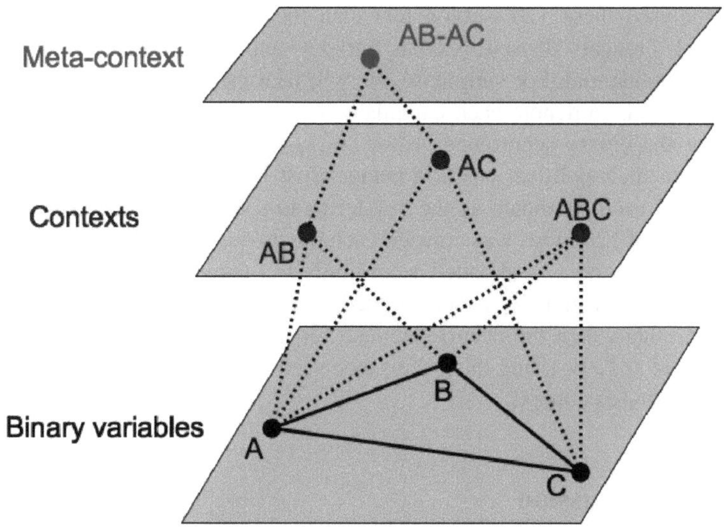

Fig. 2 Hierarchical relation between binary variables, contexts, and meta-contexts. As an example, three nodes network with binary variables A, B, and C is shown. The horizontal planes and dashed lines schematically defines the combinations to define the above level of interactions. Assume that the second order contexts AB, AC, and third order context ABC are significant. Then the meta-contexts are defined on the co-occurrence of the contexts, in which, for example, only AB-AC is significant

chy of meta^l-context up to this limit, we are able to detect all relativistic relations unconsciously associated with the text data.

4 Data Set

The campaign for the French presidential election which ended in May 2007 was marked by the strengthening of new modes of public expression. As the access to new communication media was democratizing, new patterns of public intervention have been invented. The main outcome of this new mode of intervention was that citizens could express their own views about social, political or economical topics concerning the campaign which in turn modified deeply the political agenda. The political blogosphere, in particular, is one of the main places of appropriation of public debate.

We have collected the content and hypertext links of a selection of 120 political blogs during the first six months of 2007. The database allows each blog to provide a time series of semantic attributes (quantified using a vector of occurrences of 190 key terms) that evolve according to new posts. This provides us a daily semantic characterization of each blog from January the 1st to June the 30th reflecting the current concerns of each blogger.

We have analysed these 120 weblog text data focusing on the 4 candidates (Ségolène Royal, François Bayrou, Nicolas Sarkozy, and Jean-Marie Le Pen) of the last French presidential. For simplicity, we will take the abbreviation of their family name, R,B,S,L to denote each candidate. We will use these abbreviations as subscript for the binary occurrence vector (x_R, x_B, x_S, x_L) aligned from left to right according to their political position being leftist to rightist.We set the first or family name of each candidate as the key terms to scan the weblogs and obtain the data. Each of 120 blogs was crawled each day to see whether these names appeared or not in their posts, and coded accordingly as 1 (occurrence) and 0 (non-occurrence). For example, if a blog post contained the words "Royal" and "Sarkozy" but none of the others, then the occurrence data for this blog in this day will be $(x_R, x_B, x_S, x_L) = (1,0,1,0)$. Using this occurrence vectors, we extracted significant order-wise correlation as context.

5 Result and Discussion

5.1 Context Dynamics

Both for clarity and because of lack of space, among the dynamics of all $n_c = 11$ contexts, we only focus in this paper on some interesting sets of contexts. Here, we are limited to give a case study on the above mentioned data, and further study is needed for a generalization.

One of the novelties of this method is that we are able to distinguish the significance of a context from its subcontexts. In other word, we can compare the different orders of correlation on a specific set of variables. This raise an issue about whether higher-order correlation could play an important role to characterise social dynamics. We take here only 2 opposite examples, the one where higher-order correlation is stronger (Fig.4), and the other where lower-order correlation or the subcontexts are more dominant (Fig.3). Using the abbreviation to denote the contexts, Fig.3 shows the dynamics of the 2nd order contexts RB, RS, BS and the 3rd order context RBS. While Fig.4 shows the dynamics of the 2nd order contexts RS, RL, SL and the 3rd order context RSL. Since little negative correlation is observed, only positive correlation is depicted. The distribution for each day was calculated taking the data of 7 previous days. Therefore, the significance is judged on a weekly scaling basis.

These contexts remains mostly significant during the campaign except the context RBS. The context RSL is shown to be always superior to the significance of its subcontexts. This means there was a general tendency to put the three names R,S, and L altogether instead of only two of them in blog posts. Since R and S were the two persons who actually went to the final referendum, the superiority of the context RSL suggests that L was a popular catalyst of the debate about R and S, which was actually the case in many situations.

The context RBS, on the other hand, only shows intermittent significance, and mostly behind its subcontexts. This shows the tendency of people who would rather compare each 2 persons among R,B, and S instead of discussing on these

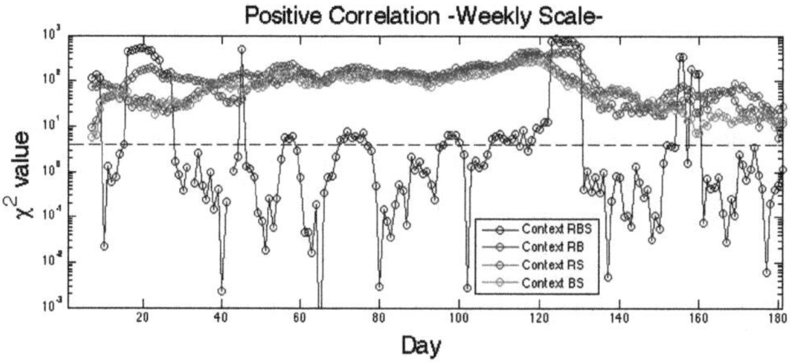

Fig. 3 Context dynamics among R, B, and S.The absolute value of λ defined in (16) is plotted in log-scale vertical axis. Only positive correlation is shown. The horizontal axis is the day counted from the beginning of the campaign. The dashed line is the 5 % significant level 3.841 of χ^2 test

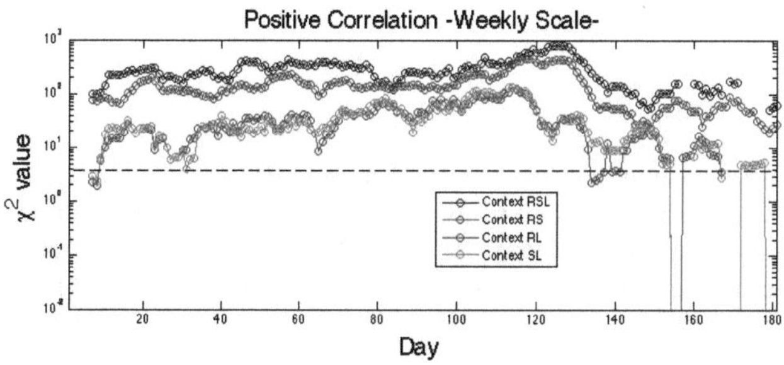

Fig. 4 Context dynamics among R, S, and L. The absolute value of λ defined in (16) is plotted in log-scale vertical axis. Only positive correlation is shown. The horizontal axis is the day counted from the beginning of the campaign. The dashed line is the 5 % significant level 3.841 of χ^2 test

3 candidates together. Since B is considered to be a moderate candidate, the inferiority of the context RBS may represent the lack of the arguments with comprehensive perspective ranging from left to right wing.

5.2 *Meta-context Dynamics*

We analyse here the 2nd-order contexts between the contexts RB, RS, BS, and RBS, or the 2nd-order meta-contexts of R,B, and S. Fig.5 shows separately the positive and negative correlation of these meta-contexts. The distribution at each day was calculated taking the data of 30 previous days. Therefore, the significance is judged on a monthly scale basis.

From this viewpoint, the whole election campaign can be characterized in 3 distinctive parts. The first 80 days are dominated by many negative correlations around 5 % significant level such as RS-BS, RB-RBS, RS-RBS, and BS-RBS. This means the significant lack of meta-context dynamics much more than an independent trial.

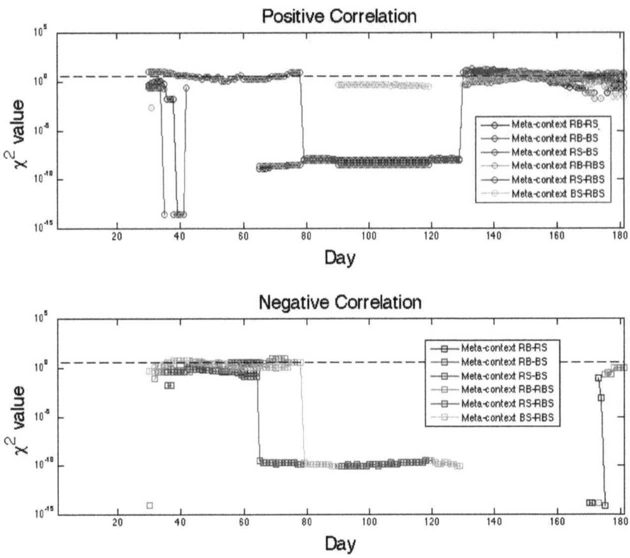

Fig. 5 Meta-context dynamics among the contexts RB, RS, BS, and RBS. The absolute value of λ defined in (16) is plotted in log-scale vertical axis. Top: Only positive correlation is shown. Bottom: Only negative correlation is shown. The horizontal axis is the day counted from the beginning of the campaign. The dashed line is the 5 % significant level 3.841 of χ^2 test

There exists however 1 constant positive correlation of RB-BS. This was then the only popular topic at meta-context level in this period. The mid term from day 80 to 130 shows the lack of significant correlation neither positive nor negative. Although the significance of contexts are rather high in this period, their meta-dynamics are absent. The last term is dominated by the 2nd-order meta-contexts around positive significance level. As the campaign get in the final third, meta-contexts highly emerged.

Possible interpretation of the meta-context in the last third may concern the second vote that comes after this campaign. In this period, people are supposed to choose two candidates who will fight the second tour, so that there should be the discussion to compare between different pairs of candidates. Therefore, the second order meta-contexts between second order contexts may reflect such arguments.

The remarkable thing is that these 3 terms are sharply separated along the time axis, and show the existence of sudden drastic change.This property is not apparent in context dynamics. This imply the importance and independence of dynamics in meta-context level, and leads us to form a hypothesis if the meta-context dynamics can serve for the global characterization in temporal scale. If the meta-context dynamics has generally such tendency to change discretely with typical transition patterns, and to be maintained for a certain period, it may allow us in real time to forecast in which phase of opinion formation the community is situated.

6 Conclusion

We have investigated all possible orders of interaction on the combinational hierarchy of 4-dimensional binary variables with the use of political weblog data, and discussed their significance. The method provided a novel way of hierarchical clustering according to the orders of interactions and its meta-level combinations, reflecting data structure into the definition of co-occurrence matrix. The defined setting provided a wide range of viewpoints, which is also applicable in general to the characterization of epistemic community from text data.

References

1. Boccaletti, S., Latora, V., Moreno, Y., Chavez, M., et al.: Complex networks: Structure and dynamics. Physics Reports (January 2006)
2. Garfield, E.: Historiographic mapping of knowledge domains literature. Journal of Information Science 30(2), 119–145 (2004)
3. Nakahara, H., Amari, S.-I.: Information-geometric measure for neural spikes. Neural Comp. 14(10), 2269–2316 (2002)
4. Strogatz, S.H.: Exploring complex networks. Nature (January 2001)
5. Wasserman, S., Faust, K.: Social network analysis (1994)
6. Holzfuss, J., Parlitz, U.: Lyapunov exponents from time series. Lecture Notes in Mathematics, vol. 1486, pp. 263–270. Springer, Heidelberg (1991)

Using Time Dependent Link Reduction to Improve the Efficiency of Topic Prediction in Co-Authorship Graphs

Susanne Hoche, David Hardcastle, and Peter Flach

Abstract. One prominent task in graphs is property prediction, where a property of some of the graph's nodes is known and used to make predictions for those individuals for which this property is unknown. In this paper, we look at topic prediction for papers organized in a so-called co-authorship graph (CAG) where the individuals are scientific papers with links between them if they share some author. A CAG tends to have a large number of cliques, each formed by all the papers published by the same author. Thus, topic prediction in a CAG tends to be computationally expensive. We investigate in how far we can reduce this complexity without sacrificing the prediction quality by reducing the number of links in the CAG based on the papers' publication dates. We apply an inexpensive iterative neighbourhood's majority vote based algorithm to predict unknown topics based on the papers with known topics and the CAG's link structure. For three data sets, we evaluate our algorithm in terms of classification accuracy and computational time on both the full graph G and subgraphs of it. On substantially smaller subgraphs of G, our algorithm obtains classification accuracies that are similar to the results obtained on G, while achieving a reduction in execution time of up to one order of magnitude.

1 Introduction

The analysis of large and complex graphs has become increasingly important in a variety of scientific disciplines. Graphs allow us to model data which consists of individuals that are connected to each other in some way. In a graph $G = (V, E)$, the individuals are modelled as nodes $v \in V$, and the connections between the

Susanne Hoche
Department of Computer Science, University of Bristol, UK
e-mail: hoche@cs.bris.ac.uk

David Hardcastle
Department of Computer Science, University of Bristol, UK
e-mail: Hardcastle.David@yahoo.co.uk

Peter Flach
Department of Computer Science, University of Bristol, UK
e-mail: Peter.Flach@bristol.ac.uk

R. Menezes et al. (Eds.): Complex Networks: CompleNet 2009, SCI 207, pp. 173–184.
springerlink.com © Springer-Verlag Berlin Heidelberg 2009

individuals as edges, or links, $e \in E$ between them. One prominent task is property prediction in a graph G, where a property of some of G's nodes is known and used to make predictions for the nodes in G for which this property is unknown.

Here we look at topic prediction for scientific papers organized in a co-authorship graph (CAG) where nodes are papers with links indicating some shared author(s). Let G be a CAG, and let G's nodes be two disjoint sets of papers with complete and missing knowledge about the topics, respectively. Our aim is to predict the unknown topics based on the known ones and G's link structure. Predicting scientific papers' topics is useful, e.g., to identify scientific communities, determine peer-reviewers or potential collaborators, and monitor the rise and fall in topics' popularity over time.

In this paper, we propose a time dependent link reduction technique to improve the efficiency of topic prediction in CAGs. A CAG tends to have a large number of cliques, each formed by all the papers published by the same author. Thus, the nodes tend to have a large number of links. We investigate in how far we can reduce the topic prediction problem's complexity without sacrificing its quality by reducing the number of links in a CAG based on the papers' publication dates.

This paper is organized as follows. We review related work in Section 2, and introduce CAGs, the task of topic prediction, and our iterative neighbourhood's majority vote based prediction algorithm in Section 3. We discuss an exponential transformation of the edge weights in a CAG to guide the reduction of the number of edges based on the papers' publication dates in Section 4. In Section 5, we present our empirical evaluation, and conclude in Section 6.

2 Related Work

Graphs induced from bibliographic data have been widely investigated (e.g., [1], [13], [14]). However, more closely related to this paper are publications in the fields of link-based object classification.

Chakrabarti et al. [2] and [15] were among the first to study the effects of using related objects' attributes, e.g., local information such as the terms occurring in a text, and external attributes, such as the known topics of linked text documents, to enhance classification in linked domains. Chakrabarti et al. [2] propose a relaxation-labelling based method for topic prediction in hyperlinked domains. Relaxation-labelling strategies, originally proposed for the field of image analysis [16], modify probability estimates by iterative adjustment, taking into account the estimates for neighbouring objects. While [2] applies a bootstrap mechanism to classify all unlabelled documents, [15] incrementally takes into account the classes of unlabelled documents only after they have been classified on the basis of neighbouring documents. However, both studies agree in that incorporating related objects' labels can enhance classification while using their local information can actually harm it.

Lafferty et al. [7] introduce conditional random fields for link-based object classification, and [17] extends this approach to a setting of arbitrary graphs instead of chains. Neville et al. [12] propose the use of relational dependency networks and Gibbs sampling to collectively infer labels for linked instances.

Lu et al. [8] propose an iterative link-based object classification method based on modelling link distributions which describe the neighbourhood of directed links around an object. The classes are iteratively updated by computing the link distributions and assigning to an object the class with the largest posterior probability. Experiments indicate that a classifier considering both the objects' local attributes and their link distributions outperforms a classifier based on local attributes only.

Macskassy et al. [9] investigate the effectiveness of relaxation labelling based relational classifiers similar to the one proposed in [2] that predict an object's classification based only on the class labels of directly linked objects. Comparative studies on different networked domains indicate that such simple relational classifiers which don't learn and consider nothing but known class labels of related entities can serve as a baseline comparison for relational learning.

Link-based object classification has been an important research topic over the past years and as a result, numerous relaxation labelling based methods exist. Our method is similar to one of the methods discussed in [9]. However, [9] iterate their method a standard 100 times which makes it very expensive, especially in the presence of large and complex graphs. There also exist differences in the way the unknown nodes' estimates are considered for the predictions' update.

3 Predicting Paper Topics in a Co-Authorship Graph

In a co-authorship graph G, nodes are papers which are linked to each other by an undirected edge if and only if the papers have an author in common. $G = (V,E)$ consists of a set of nodes $V = \{v_1,\ldots,v_n\}$ and a set $E = \{(v_i,v_j)\}$ of undirected edges between $v_i,v_j \in V$ which we assume to be weighted. We define the weight w_{ij} of the link between two papers v_i and v_j as

$$w_{ij} := \begin{cases} \sum_{i=1}^{n} \frac{1}{a_i} & \text{if } v_i,\ v_j \text{ have overlapping authors } A_1,\ldots,A_n, n \geq 1 \\ 0 & \text{otherwise,} \end{cases} \tag{1}$$

where a_i is the total number of papers that author A_i has published (in G). This definition is in the style of Mark Newman's definition of collaboration strength between authors [14] and expresses that each author contributes the same strength to the weight of each pair of papers (co-)authored by her in G. In our previous experiments, this definition has been proven most successful for encoding edge weights.

A node's *target attribute* which we aim to predict can take on a certain *class label*, or *topic*, $\phi \in \Phi$ which is partially observed for the nodes in V. Note that a paper can belong to several topics. Let $V = V^K \cup V^U, V^K \cap V^U = \emptyset$, where V^K is the set of *known nodes* with known target attributes, and V^U is the set of *unknown nodes* with unknown target attributes. The task is to predict – on the basis of V^K and G's link structure given by E – for each topic $\phi \in \Phi$ and each node $v \in V^U$, v's probability of exhibiting ϕ. To this end, we apply our iterative neighbourhood's majority vote prediction algorithm ("iNMV"). It is based on a simple majority vote of directly linked nodes and consists of an *initialisation step* and an *update step* which can be applied iteratively. Initially, we assign to each unknown node for each topic $\phi \in \Phi$ an estimate on the basis of observing ϕ for the known nodes. This estimate is iteratively modified based on the neighbouring nodes' estimates for ϕ.

iNMV learns from positive, negative and unlabelled examples, where for each topic $\phi \in \Phi$, the nodes $v \in V^K$ with topic ϕ constitute the positive examples for ϕ, those nodes in $v \in V^K$ which do not have topic ϕ constitute the negative examples for ϕ, and the unknown nodes V^U constitute the set of unlabelled examples. iNMV derives for each topic $\phi \in \Phi$ and each unknown node $v_i \in V^U$, a class membership estimate expressing the probability of observing ϕ for v_i.

Our approach assumes that nodes in the same neighbourhood of the graph tend to have similar topics, and that the predicted topic for one node in the graph depends on the topic of its direct neighbours. Therefore, we assume that the probability of observing topic ϕ for node $v_i \in V^U$ given G is equal to the probability of observing ϕ for v_i given v_i's neighbourhood $N_i := \{v_j \in V|(v_i, v_j) \in E\}$.

The class membership estimate for an unlabelled node v is based on the labels of the nodes in V^K that are directly linked to v and on the class membership estimates of the nodes in V^U directly linked to v. Thus the classification of an entity depends on the classification of all neighbouring entities. This way, entities are classified in dependence of each other, and mutual influence of the predictions is accounted for.

In order to predict the probability of observing topic ϕ_k for a node $v_i \in V^U$ with unknown target attribute y_i, we assign to v_i an initial probability estimate $p_{ik}^{(1)} := P(y_i = \phi_k|V^K)$, where $P(y_i = \phi_k|V^K)$ denotes the probability of observing ϕ_k for v_i conditioned on the topics observed for the nodes in V^K. Note that this is done separately and independently for all $\phi_k \in \Phi$.

This initial probability is estimated based on the number n_k of times that ϕ_k is observed in V^K using the maximum likelihood based m-estimate where the observations are augmented by m additional samples which are assumed to be distributed according to p:

$$p_{ik}^{(1)} := P(y_i = \phi_k|V^K) = \frac{n_k + p \cdot m}{|V^K| + m}, \tag{2}$$

where $|V^K|$ denotes the cardinality of the set V^K. We choose $m = 1$ and $p = 0.5$ (each topic is equally likely to be present or absent). For a node $v_i \in V^K$ with known target attribute y_i, let $p_{ik}^{(1)} := 1$ for every topic ϕ_k that is observed for v_i (a paper can have more than one topic).

For each topic $\phi_k \in \Phi$, we update the initial probability estimates $p_{ik}^{(1)}$ for each node $v_i \in V^U$ based on the observations and estimates for ϕ in v_i's neighbourhood. Separately and independently for each $\phi_k \in \Phi$, the modified estimate $p_{ik}^{(t+1)} := P^{(t+1)}(y_i = \phi_k|N_i)$ is derived on the basis of the estimates $p_{jk}^{(t)} := P^{(t)}(y_j = \phi_k|N_j)$ for observing ϕ_k for v_i's neighbours $v_j \in N_i$ in the t-th update step

$$p_{ik}^{(t+1)} := P^{(t+1)}(y_i = \phi_k|N_i) = \frac{1}{\sum_{n_j \in N_i} w_{ij}} \sum_{n_j \in N_i} w_{ij} p_{jk}^{(t)}, \tag{3}$$

with w_{ij} defined according to (1).

For an undirected graph which we deal with here, (3) is recursive. To account for the mutual influence between linked nodes, the estimates can be propagated through the graph by iterating (3) several times. The more often the update step is iterated the further are the predictions propagated through the graph.

4 Time Dependent Link Reduction

We investigate the effect of time dependent link reduction on topic prediction in a CAG in terms of prediction quality and complexity. Since a CAG tends to have a high degree of cliques, its nodes tend to have a large number of links which makes topic prediction computationally expensive. We investigate how far we can reduce the prediction's complexity and how much the quality of topic prediction in G is affected when reducing the number of links in G on the basis of the papers' publication dates.

To this end, we apply an exponential transformation to the edge weights in G such that the original weight w_{ij} of an edge between two papers p_i and p_j in G decreases exponentially with the absolute difference $|year_i - year_j|$ between the publication years $year_i$ and $year_j$ of p_i and p_j. Let the adjacency matrix A contain for each node pair $(v_i, v_j) \in E$ an entry $A_{ij} = w_{ij}$ with w_{ij} defined according to (1). We then apply the exponential transformation

$$A_{ij} \mapsto A_{ij} \cdot 2^{-|year_i - year_j|}. \tag{4}$$

Let A_t be the adjacency matrix resulting from this transformation of A which assigns exponentially decreasing weights to links between papers as the absolute difference between their time of publication gets larger. This way, edges between papers with similar publication dates get relatively more weight than links between papers that are more distant with respect to their publication dates.

In a second step, we delete n percent, $n = 25, 35, 50, 60, 70, 80, 90$, of the edges in A_t with the lowest transformed edge weights plus all edges in A_t with a weight equal to the largest weight of any edge such removed, resulting in the adjacency matrix $A_{t,n}$. In what follows, we will refer to this as deleting n^+ percent of the edges in A_t.

Additionally, we generate the adjacency matrices A_n by deleting the same edges as in $A_{t,n}$ but then restoring the original edge weights from A for the retained edges. In what follows, we will refer to the edge weights in $A_{t,n}$ and A_n as the *transformed weights* and the *original weights*, respectively.

5 Empirical Evaluation

We evaluate the effects of the link weighting and removal described in Section 4 on our iNMV method in terms of prediction quality and efficiency on CAGs induced from the bibliographic data sets "IPLNet2" [6] and "Cora" [10].

5.1 Data

ILPNet2 is a bibliographic database of hand selected references related to Inductive Logic Programming. Cora is a collection of computer science research papers that has been automatically collected from the web [10]. From each data sets, we extract the largest connected component. It consists of 406 nodes with known topics and 6325 undirected links (on average ≈ 15 links per node) for ILPNet2, and of 10,513 nodes with known topics and 87,438 undirected links (on average ≈ 8 links per node) for Cora. We restrict our evaluation to the 10 topics that include at least 20 papers each for ILPNet2, and to the 6 top-level topics with the highest number of positive examples ("6 Top"), and the 7 Machine Learning sub-topics on the lowest level of the topic hierarchy ("7 ML") for Cora.

For each topic $\phi \in \Phi$, we generate in t trials and an f-fold cross validation manner $p = t \cdot f$ distinct training and test set partitions, using stratified sampling, i.e. the ratio of positive and negative examples in all training and test sets is the same as in the original data set, with $t = 10, f = 4$ for ILPNet2, and $t = 5, f = 2$ for Cora "6 Top" and "7ML".

We generate $u\%$ unknown nodes by deleting the class label indicating to which topic(s) a paper belongs, but use the label, after we have obtained iNMV's prediction, to evaluate iNMV's performance in terms of Receiver Operating Characteristics (ROC) analysis (see Section 5.2 below). We choose $u = 25$ for ILPNet2, and $u = 50$ for Cora. For each topic ϕ, the class label is deleted uniformly at random from $u\%$ of the publications belonging and not belonging to ϕ, respectively. This way, each such partition is a sample stratified with respect to ϕ, reflecting the distribution of topic ϕ in the original data set. The nodes with removed labels are distinct in each sample, and their union results in the complete set of publications. Thus, each publication serves for each topic exactly once as an unlabelled example.

For each topic ϕ and each node $v \in V^U$, we yield from each of the p partitions an estimated degree to which v belongs to ϕ. These values can be interpreted as scores which we use to rank the nodes, where a higher score indicates a higher probability of exhibiting ϕ. For each topic ϕ, we average the results over all trials and partitions, and evaluate the resulting rankings of the unknown nodes in terms of ROC analysis.

5.2 Evaluation

Receiver Operating Characteristics (ROC) analysis is widely used to depict for binary classifiers the tradeoff between their sensitivity, the proportion of correctly classified positives, and specificity, the proportion of correctly classified negatives [4]. The area under the ROC Curve (AUC) statistic is the most important statistic associated with ROC analysis. It is a single scalar value representing a classifier's expected performance. It can be interpreted as the probability that the classifier will assign for a tuple $(+,-)$ of a positive and a negative example drawn uniformly at random, a higher score to the positive example than to the negative (which means that these two examples are ranked correctly relative to each other). A classifier's

AUC is the fraction of $(+,-)$-tuples that it correctly ranks relative to each other, and is defined as

$$AUC = \frac{\sum_{i=1}^{m} \sum_{j=1}^{n} 1_{(+_i > -_j)}}{m \cdot n}, \tag{5}$$

where $+_1, \ldots, +_m$ are the scores assigned to the m positive examples, $-_1, \ldots, -_n$ are the scores assigned to the n negative examples, and $1_{(+_i > -_j)}$ is the indicator function which is equal to 1 if $+_i > -_j$, and 0 otherwise. A classifier's AUC is maximal, i.e. equal to 1, if and only if it ranks all positive examples higher than the negative examples. Any misranked $(+,-)$-tuple decreases the AUC.

5.3 Experimental Setup

For each data set we apply iNMV with 1, 5 and 20 iterations of the update step to the matrices $A_{t,n}$ and A_n with n^+ percent of the edges with lowest transformed weights removed for $n = 0, 25, 35, 50, 60, 70, 80, 90$, and compare the resulting AUCs and run times (measured on an Intel(R) Xeon(TM) MP CPU 3.16GHz processor). We focus our discussion of the results on the following questions: for which percentage n^+ and for which number of iterations do we obtain the highest AUCs and lowest run times, respectively, and do transformed or original edge weights perform better.

In a control experiment, we delete n^+ percent of the links uniformly at random to exclude a purely random effect of time based link reduction.

We also perform a significance test. When comparing more than two classifiers, the Friedman test [5] is widely recommended [3] which compares k algorithms over N data sets by ranking each algorithm on each data set separately. If the null-hypothesis - that all algorithms are performing equivalently - is rejected under the Friedman test statistic, post-hoc tests such as the Nemenyi test [11] can determine which algorithms perform statistically different. All results of the Friedman test reported in this paper are obtained for a 95% confidence level. Note, that for each topic ϕ, distinct partitions are generated, and that we obtain for each domain $N = |\phi|$ mutually independent data sets after averaging over all partitions and trials.

5.4 Results

Figure 1 shows boxplots of the distribution over topics of the average AUCs. For each topic, the AUC is averaged over all trials and partitions. For each CAG, we show the AUCs obtained from iNMV with $k = 1, 5, 20$ iterations ("Its") on $A_{t,n}$ and A_n for an increasing fraction n^+ of removed edges. Each boxplot shows the median, lower and upper quartile, and the lower and upper limit of the AUCs for the single topics.

On the **ILPNet2** CAG, decreasing the edge weights exponentially with respect to the difference between linked papers' publication years and deleting n^+ percent of the edges with the lowest transformed weights seems to be more advantageous – in

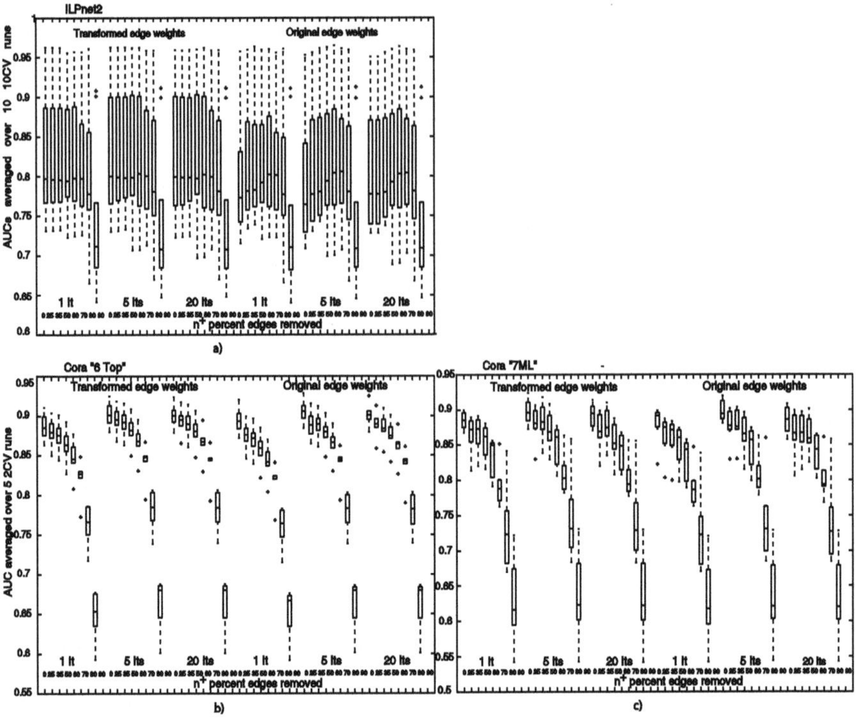

Fig. 1 Boxplots showing the distribution over topics of the average AUCs. For each topic, the AUC is averaged over all trials and partitions. The results are shown for various combinations of n^+ (the percentage of edges with lowest transformed weight deleted) and the number of iterations k, for ILPNet2 (a), Cora "6 Top" (b) and Cora "7 ML" (c). We show - from left to right - the results for $n = 0, 25, 35, 50, 60, 70, 80, 90$ for iNMV with transformed and original weights, respectively, and with $k = 1, 5, 20$ iterations ("Its"). Each boxplot shows the median, lower and upper quartile, and the lower and upper limit of the data points (not considered to be outliers), i.e. the AUCs for the single topics, for one specific n and k. An outlier is depicted as "+"

terms of AUC – than using all edges as long as n is not too large. However, according to the Friedman and Nemenyi post-hoc test, statistically significant differences exist, for any k, only between the AUCs obtained for $n = 0, \ldots, 60$ and $n = 90$.

Either 1 or 5 iterations of iNMV's update step predominantly result in the best AUCs (for $k = 5$ 75% of the time for transformed weights, and for $k = 1$ 62.5% of the time for original weights), but with no statistical significance.

Transformed weights result for all n but $n = 80$ in larger AUCs than original weights, although without statistical significance. This indicates that there is extra information in the publication dates that can be exploited for the label prediction.

For both transformed and original edge weights, the median of the AUCs over all trials and partitions for iNMV with $k = 1, 5, 20$ iterations on the ILPNet2 data set is maximal when deleting $n^+ = 60.3\%$ of the links which results in deleting almost all links between papers published 4 or more years apart.

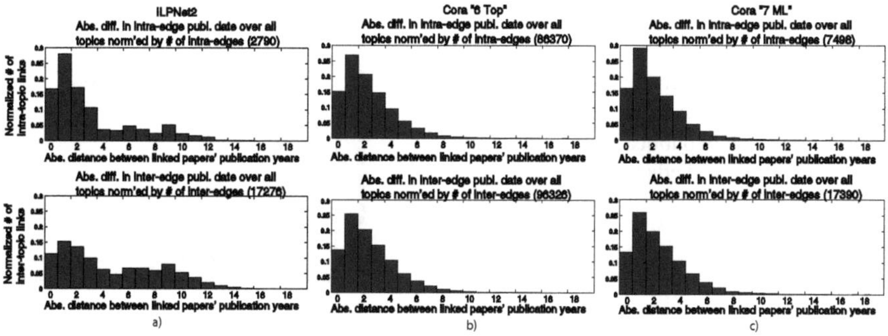

Fig. 2 Histograms of intra- and inter-topic edges' distance with respect to linked papers' publication years for ILPNet2 (a) and Cora "6 Top" (b) and Cora "7 ML" (c)

Figure 2 a)-c) shows for ILPNet2, Cora "6Top" and Cora "7 ML" the distribution of linked papers that are on the same topic (intra-topic links) and on different topics (inter-topic links). Summing up the first four bins of the upper and lower histogram in Figure 2 a), i.e. pairs of papers that are no more than 3 years apart, we see that they comprise about 72% of all intra-topic links and about 50% of all inter-topic links of the ILPNet2 CAG. Thus, 72% of the links between papers with the same topic are retained when removing links between papers 4 or more years apart, but only 50% of the links between papers on different topics. Figure 3 shows, that this difference $72-50$ is the maximum difference between the fraction of intra- and inter-topic edges retained for any fraction n^+ of removed edges.

For both original and transformed edge weights, topic prediction with iNMV with $k = 1,5,20$ iterations in the ILPNet2 CAG is robust against the deletion of n^+ percent of the links with the lowest transformed weights for up to $n^+ = 81$, which means that AUCs can be achieved which are not significantly lower than those obtained on the full graph even if up to 81% of the links are ignored.

On the **Cora** CAGs, decreasing the edge weights exponentially with respect to the difference between linked papers' publication years and deleting n^+ percent of the edges with the lowest transformed weights leads for $k = 1,5,20$ and both transformed and original weights to a steady decay of AUCs as n increases. According to the Friedman and Nemenyi post-hoc test, there are statistically significant differences between the AUCs obtained for $n = 0,\ldots,35$ and $n = 70,\ldots,90$ for Cora "6 Top", and between the AUCs obtained for $n = 0,\ldots,50$ and $n = 70,\ldots,90$ for Cora "7 ML" for $k = 1,5,20$.

Here, 5 iterations of iNMV's update step generally result in the best AUCs for transformed and original weights, although with no statistical significance. Using transformed weights results for $k = 1,5,20$ in 83% of the time for Cora "6 Top" and 54.2% of the time for Cora "7 ML" in larger AUCs than using the original weights, although without statistical significance. Again, this indicates that additional information in the publication dates can be used to enhance topic prediction.

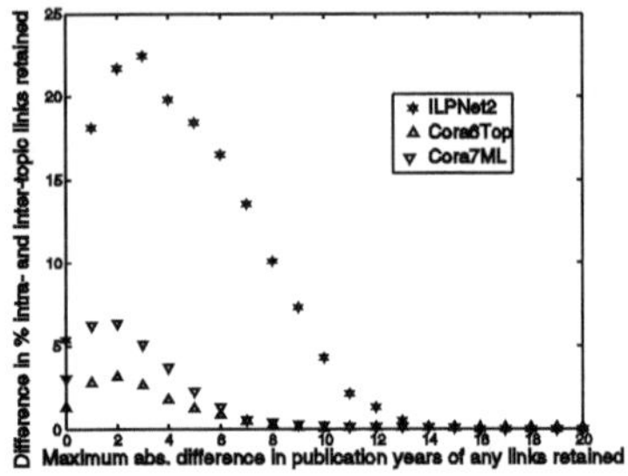

Fig. 3 Difference between percent intra- and inter-edges retained vs. maximum absolute distance in publication years of retained edges

From Figures 2 b) and c) and 3, we see that the intra- and inter-topic edges for Cora cannot be as nicely separated by using our exponential transformation as for ILPNet2. The maximum difference between intra- and inter-topic edges retained for both Cora data sets exists for a maximal absolute difference of 2 years in linked papers' publication dates (cf. Figure 3). However, this difference is in both cases only around 5%, which means that proportionally as many intra-topic as inter-topic links are retained when deleting all edges between papers more than 2 years apart. There are 52.7% and 69.9% inter-topic links, respectively, on the full Cora CAGs. Deleting n^+ percent of the inter-topic edges results for any n in removing proportionally as many intra-links which does not alleviate the prevalence of inter-topic edges. Thus, for the Cora data sets, there is no benefit – in terms of AUC – of removing links in the co-authorship graph. However, for any k, AUCs do not significantly decrease up to a removal of $60^+\%$ of the links for both Cora data sets.

ILPNet2 and Cora For all CAGs, random deletion of links leads, as opposed to link weighting and selective removal, to an increased number of isolated nodes and a significant decrease of AUC. The publication date holds information about papers' similarity and is thus a valid guide for deleting edges in the graph.

For all CAGs, selective removal of $n^+\%$ of the links results in an increased reduction in run time for increasing n. While this reduction is rather small for the small ILPNet2 CAG, viz. a reduction by $1.95\%, 6.2\%, 11.46\%$ for $k = 1, 5, 20$ iterations of iNMV's update step, respectively, and a removal of $80^+\%$ of the links, it is rather substantial for the two Cora CAGs: removing $60^+\%$ of the links leads to a reduction of run time of $12.6\%, 34.5\%, 62.1\%$ for Cora "5 Top", and $17.1\%, 37.4\%, 67.3\%$ for Cora "7 ML", for $k = 1, 5, 20$ iterations, respectively.

5.5 *Summary and Implications*

For all domains, removal of $n^+\%$ edges in the CAG based on the papers' publication dates results only in a statistically non-significant decrease of AUC up to large n – unlike random edge deletion. Moreover, the run time complexity decreases, rather substantially for the large CAGs, with increasing n.

However, not all CAGs are equally suited for this type of link reduction. For ILPNet2, we can determine an absolute difference δ between papers' publication years for which we obtain a ratio of removed intra- and inter-topic links optimal for topic prediction. However, we do not obtain a ratio of removed intra- and inter-topic links optimal for topic prediction for such δ for either of the Cora CAGs.

A most likely explanation for this discrepancy is the difference in the data sets' origins. ILPNet2 is a hand selected collection of papers in a rather specific scientific discipline, and most authors' papers from the years considered will be exhaustively included in this data set. The histograms in Figure 2 a) show that the range of publication dates of authors may span well up to 15 years. However, our investigation also indicates that the topics in the ILPNet2 data set are rather time dependent, i.e., if authors collaborate over a longer period of time (more than 4 years), the scientific field they are collaborating in is likely to change.

By contrast, the Cora data set is web crawled and does most likely not contain an author's complete scientific work for the considered time span (which is also suggested by the fact that a Cora paper has on average about 50% of the neighbours that an ILPNet2 paper has). Moreover, its topics are much more diverse, and while many ILPNet2 papers are labelled by the people who collected the data, the Cora topics have been applied by the authors themselves. The histograms in Figure 2 b) and c) show that the authors' publication dates span up to around 10 years, with no clear indication of topic coherence.

These observations may suggest to investigate a CAG's properties such as the ratio of intra- and inter-topic links in dependence of distance in publication years, and then select a suitable fraction n of links to be removed, and a suitable number k of the prediction algorithm's update steps.

Over all, the results indicate that there is no need to use all links in a CAG, since a large link reduction leads to only non-significantly lower AUCs and at the same time to a substantial decrease in run time complexity.

6 Conclusion

We have investigated the effect of time dependent link reduction on the efficiency and quality of topic prediction for papers in a co-authorship graph G with partially observed topics. To this end, we have applied an exponential transformation of the edge weights in G to guide the removal of a fraction of edges in G based on the papers' publication dates. Up to a large number of removed edges, topic prediction with our iNMV algorithm results in AUCs that are not significantly lower than those obtained on the full CAG, while substantially reducing the run time complexity. Our empirical evaluation suggests that the different origins of bibliographic collections render it necessary to select specific parameters for both the edge removal and the prediction algorithm depending on the specific nature of the induced CAG.

Acknowledgements. The authors would like to acknowledge funding and support for this work from GCHQ in Cheltenham in the UK.

References

1. Barabasi, A., Jeong, H., Neda, Z., Ravasz, E., Schubert, A., Vicsek, T.: Evolution of the social network of scientific collaborations. Physica A 311(3-4), 590–614 (2002)
2. Chakrabarti, S., Dom, B.E., Indyk, P.: Enhanced hupertext categorization using hyperlinks. In: Proc. of SIGMOD 1998, ACM Int. Conf. on Management of Data, pp. 307–318 (1998)
3. Demšar, J.: Statistical comparisons of classifiers over multiple data sets. Journal of Machine Learning Research 7, 1–30 (2006)
4. Fawcett, T.: An introduction to roc analysis. Pattern Recognition Letters 27(8), 861–874 (2006)
5. Friedman, M.: The use of ranks to avoid the assumption of normality implicit in the analysis of variance. J. of American Statistical Association 32, 675–701 (1937)
6. ILPNet2, `http://www.cs.bris.ac.uk/~ilpnet2/tools/reports`
7. Lafferty, J., McCallum, A., Pereira, F.: Conditional random fields: Probabilistic models for segmenting and labeling sequence data. In: Proc. 18th International Conf. on Machine Learning, pp. 282–289 (2001)
8. Lu, Q., Getoor, L.: Link based classification. In: Proceedings of the International Conference on Machine Learning, pp. 496–503 (2003)
9. Macskassy, S.A., Provost, F.: Classification in networked data: A toolkit and a univariate case study. Journal of Machine Learning 8, 935–983 (2007)
10. McCallum, A., Nigam, K., Rennie, J., Seymore, K.: Automating the construction of internet portals with machine learning. Information Retrieval 3(2), 127–163 (2000)
11. Nemenyi, P.B.: Distribution-free multiple comparisons. PhD thesis, Princeton University (1963)
12. Neville, J., Jensen, D.: Iterative classification in relational data. In: Proc. AAAI 2000 Workshop on Learning Statistical Models from Relational Data, pp. 13–20 (2000)
13. Newman, M.: The structure of scientific collaboration networks. In: Proceedings of the National Academy of Sciences, vol. 98, pp. 404–409 (2001)
14. Newman, M.: Coauthorship networks and patterns of scientific collaboration. In: Proceedings of the National Academy of Sciences 101, pp. 5200–5205 (2004)
15. Oh, H.-J., Myaeng, S.H., Lee, M.-H.: A practical hypertext categorization method using links and incrementally available class information. In: Proc. of SIGIR 2000 (2000)
16. Rosenfeld, A., Hummel, R., Zucker, S.: Scene labeling by relaxation operations. IEEE Transactions on Systems, Man and Cybernetics 6, 420–433 (1976)
17. Taskar, B., Abbeel, P., Koller, D.: Discriminative probabilistic models for relational data. In: Proc. of UAI 2002, pp. 485–492 (2002)

Fast Similarity Search in Small-World Networks

Kazuo Aoyama, Kazumi Saito, Takeshi Yamada, and Naonori Ueda

Abstract. We present a novel graph-based approach for fast similarity searches suitable for large-scale and high-dimensional data sets. We focus on a well-known feature of small-world networks, they are "searchable," and propose an efficient index structure called a degree-reduced nearest neighbor graph. A similarity search is then formulated as a problem of finding the most similar object to a query object by following the links in this graph with a best-first neighborhood search algorithm. The experimental results show that the proposed search method significantly reduces search costs. In particular, we apply it to data sets consisting of nearly one million documents, and successfully reduce the average number of similarity evaluations to only 0.9% of the total number of documents.

1 Introduction

Considerable attention has been devoted in recent years to investigating the small-world phenomenon in social networks: any two individuals in a network are likely to be connected through short chains of intermediate acquaintances [1]. From an algorithmic viewpoint, the most striking feature is that small-world networks are "searchable," i.e., individual nodes using only local information are collectively very effective at constructing short paths between any two nodes in the network [2,3]. As shown in Milgram's experiment [4], ordinary people can deliver messages through their network of acquaintances to reach a specific but distant target person in only a few steps.

Kazuo Aoyama, Takeshi Yamada, and Naonori Ueda
NTT Communication Science Laboratories, Kyoto, Japan
e-mail: `issei@cslab.kecl.ntt.co.jp`

Kazumi Saito
University of Shizuoka, Shizuoka, Japan
e-mail: `k-saito@u-shizuoka-ken.ac.jp`

R. Menezes et al. (Eds.): Complex Networks: CompleNet 2009, SCI 207, pp. 185–196.
springerlink.com © Springer-Verlag Berlin Heidelberg 2009

Instead of exploring the properties of real social networks, in this paper, we exploit this remarkable feature for developing new efficient engineering applications. In particular, we focus on a similarity search problem [5] as an application, i.e., for a given query object, the problem is to effectively find the most similar object in the database. We connect similar and closely related objects that satisfy certain conditions by links and construct a similarity network of the objects in the database, which resembles the acquaintance networks of people.

By treating the most similar object to the query as the unknown target person, we can formulate a similarity search as a message passing problem. Starting from an arbitrarily selected source object node, individual nodes pass a message to the locally connected node that is the most similar to the given query object. Here, we need to construct an efficient "searchable" similarity network of objects.

This paper proposes a degree-reduced nearest neighbor (*DRNN*) graph as an index structure of a graph-based similarity search method. The index structure enables us to quickly find the most similar object to a given query object by iteratively following links in the graph with a best-first manner.

The remainder of this paper is organized as follows: Section 2 refers to related work. Section 3 provides preliminaries required to understand the proposed search method. Section 4 details a novel algorithm for constructing a "searchable" similarity network. Section 5 presents the experimental settings and results and shows the effectiveness of the proposed method. The last section gives our conclusion.

2 Related Work

The small-world phenomenon has been extensively studied in the context of network analysis research. Kleinberg proved that a simple greedy strategy, which scans all neighboring nodes and selects the closest to the target, could efficiently pass a message when the nodes are situated on an m-dimensional lattice and additional connections are placed between any two nodes with probability proportional to r^{-m}, where r is the Euclidean distance between them [6]. Watts-Dodds-Newman [3], who considered a hierarchical case where individuals belong to groups embedded hierarchically into larger groups, also showed that a simple greedy strategy is efficient. These studies shed light on the theoretical aspects but did not consider the engineering implication of their models.

The *directed* k-nearest neighbor graph (directed *k-NN* graph for short) is a graph in which nodes p and q are connected by a directed link (edge) if the distance between p and q is within kth smallest among the distances from p to other objects. It is often used in similarity searches. For example, both Orchard [7] and Sebastian-Kimia [8] utilized this type of graph with some extensions and employed simple greedy search algorithms. However, the efficiency of a simple greedy search on a directed *k-NN* graph is limited, and the search performance strongly depends on the choice of initial nodes. Moreover, the greedy search algorithm requires that the graph is strongly connected to reach any node by traversing along the directed links.

Searching for a target node in a network is related to the decentralized broadcast mechanism in peer-to-peer networks such as Gnutella [9], because search efficiency is crucial for both problems. However, the main difference between decentralized peer-to-peer broadcasting and the (possibly more centralized) similarity search is that avoiding network overload is another compelling issue for peer-to-peer broadcasting. A different type of peer-to-peer network, which is inspired by social acquaintance networks, is described in [10, 11].

Locality-Sensitive Hashing (LSH) is a powerful dimensionality reduction method based on probabilistic hashing [12]. While LSH provides efficient nearest neighbor search, it requires each object to be represented in some specific form such as a vector in an Euclidean space, a binary vector, or a permutation of enumerative items. In contrast, it is only a similarity function defined between objects that our proposed method requires.

Bustos-Navarro-Chávez proposed a pivot-based similarity search in which a set of objects is selected as a pivot set [13]. The number of online distance calculations is efficiently reduced by utilizing the triangle inequality and pre-calculated pivot-object distances. However, the triangle inequality cannot efficiently reduce the distance calculations in a high-dimensional metric space due to the *curse of dimensionality* [5].

3 Preliminaries

Let $\mathscr{X}$ be the space of objects, where similarity function $\rho(x,y)$ is defined for any two objects $x,y \in \mathscr{X}$ that satisfies $\rho(x,y) = \rho(y,x) \in [0,1]$ and $\rho(x,y) = 1 \Leftrightarrow x = y$. Let $X \subset \mathscr{X}$ be a set of objects stored in a database and let $q \in \mathscr{X}$ be a query object, not necessarily in X. A set of the most similar objects to q in the database is expressed by $R(q) = \arg\max_{x \in X} \{\rho(x,q)\}$. Given a query object (*query* for short) and a set of objects X, we consider the search problem of finding at least one object $x^* \in R(q)$.

Our approach deals with similarity search as a network (graph) search problem: the problem of a neighborhood search that is equivalent to a message passing problem on a network (graph). We consider particular kinds of networks (graphs), namely, undirected neighborhood networks (graphs), where nodes (vertices) correspond to objects in X and links (edges) are generated between the objects that satisfy certain conditions based on similarity function ρ. The graph search problem is to find at least one vertex $x^* \in R(q)$ by iteratively following the edges in the graph starting from initial vertex $x_0 \in X$ with a neighborhood search algorithm.

3.1 Undirected k-Nearest Neighbor Graph

We start from the following simple *undirected* k-nearest neighbor graph Γ_k as a baseline. Hereafter, the graphs we consider are all undirected unless otherwise stated. Let $N_k(x)$ be a set of the top k similar objects from $x \in X$. Then undirected *k-NN* graph Γ_k is a graph with X as the set of vertices and with undirected edges between $x \in X$

and each element in $N_k(x)$. Let $V_k(x)$ be a set of the adjacent vertices of x directly connected with the undirected edges in Γ_k and $E_k(x) = \{(x,y) : y \in V_k(x)\}$ a set of the undirected edges. A simple algorithm to construct Γ_k, which we call the *nearest neighbor* (*NN*) algorithm, is shown in Algorithm 1.

It is obvious that Γ_k naturally exhibits *homophily*: the tendency of *like to associate with like*, it is likely that Γ_k also exhibits *degree disparity*: some nodes have much higher degree than others because of the undirected connections, especially when the dimensionality of the objects is high. These two characteristics are important in supporting that a small-world network is "searchable" [14].

Algorithm 1. *NN Algorithm*

Input: X
Output: Γ_k
1. **for all** $x \in X$ **do**
2. Set $V_k(x) \leftarrow N_k(x) \cup \{y : x \in N_k(y)\}$
3. **end for**
4. Set $\Gamma_k \leftarrow (X, \cup_{x \in X} E_k(x))$, where $E_k(x) = \{(x,y) : y \in V_k(x)\}$

3.2 Best-First Neighborhood Search Algorithm

We describe a best-first neighborhood search (*BS*) algorithm on given graph Γ shown in **Algorithm** 2. Here, the inputs to *BS* algorithm $BS(q, x_0, \Gamma, \beta)$ are query $q \in \mathscr{X}$, initial vertex x_0, graph Γ such as the *k-NN* graph Γ_k described above and a *k-DRNN* graph described below, and upper bound β on the search cost measured by the number of similarity evaluations. It returns most similar vertex z within bound β. Upper bound β is necessary because most Web applications for similarity searches need to guarantee a short-time response.

In **Algorithm** 2, A is the set of *evaluated* vertices whose similarity to query q is already calculated, $B \subset A$ is the set of *closed* vertices whose neighbor vertices are already evaluated, and $A \setminus B$ is the set of *open* vertices. Here, $|A|$ denotes the number of elements in A and thus represents the number of similarity evaluations.

Algorithm 2. *BS Algorithm*: $BS(q, x_0, \Gamma, \beta)$

Input: $q \in \mathscr{X}, x_0 \in X, \Gamma, \beta$
Output: z
1. Set $A \leftarrow \{x_0\}, B \leftarrow \emptyset, x \leftarrow x_0, z \leftarrow x$ // $\rho(x_0, q)$ is computed.
2. **while** $(|A| < \beta) \wedge (\rho(z,q) < 1)$ **do**
3. Set $A \leftarrow A \cup V(x), B \leftarrow B \cup \{x\}$
4. Set $x \leftarrow \arg\max_{y \in A \setminus B} \{\rho(y,q)\}$ // $\rho(y,q)$ is cached and computed only once.
5. **if** $\rho(x,q) > \rho(z,q)$ **then**
6. Set $z \leftarrow x$
7. **end if**
8. **end while**

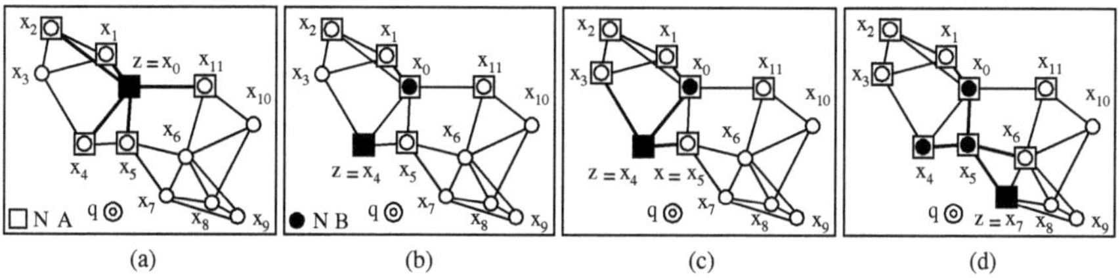

Fig. 1 The best-first neighborhood search algorithm applied to 3-*NN* graph Γ_3

If we replace lines 4–7 by "$z = x \leftarrow \arg\max_{y \in V(x)} \{\rho(y,q)\}$," then we have a version of greedy searches mainly used in the small-world network analysis. However, in this paper, we prefer a best-first search for better search efficiency.

Figure 1 illustrates how the *BS* algorithm works when applied to small 3-*NN* graph Γ_3 consisting of 12 vertices with $\beta = 8$. At each step, the sets A and B change, and the vertices in A and B, and vertex z are represented with squares (□), closed circles (●), and closed square (■), respectively. Query q is marked as a double circle (◎). Figure 1(a) shows the initial state where $z \leftarrow x_0$, $A = \{x_0, x_1, x_2, x_4, x_5, x_{11}\}$, and $B = \{x_0\}$. Figure 1(b) shows the state at the end of the first round of the **while loop** and z is updated as x_4 based on lines 4–7 in **Algorithm** 2. Figure 1(c) shows the state at the end of the second round of the **while loop** where $A = \{x_0, x_1, x_2, x_3, x_4, x_5, x_{11}\}$, $B = \{x_0, x_4\}$, $x \leftarrow x_5$, and z remains unchanged. Figure 1(d) shows the final state at the end of the third round of the **while loop**, where $A = \{x_0, x_1, x_2, x_3, x_4, x_5, x_6, x_7, x_{11}\}$, $B = \{x_0, x_4, x_5\}$, and z is updated as x_7 At this point, $|A| = 9 > \beta$ and **Algorithm** 2 returns x_7 as the most similar vertex to q. To summarize, B is updated by including vertices $x_0 \to x_4 \to x_5$ in this order, and z is updated as $x_0 \to x_4 \to x_7$.

3.3 Search Performance Measures

The solution obtained by the *BS* algorithm is not an exact but an *approximate* nearest neighbor object to given query q, just as in most conventional graph-based search methods including Orchard [7] and Sebastian-Kimia [8]. Therefore, we need to evaluate the approximation accuracy.

When we perform a best-first search using the *BS* algorithm on *k*-*NN* graph Γ_k, we have to determine parameters k and β that are related to the graph structure and the search cost, respectively. We introduce two measures, *success ratio* and *expected search cost*, to evaluate how these parameters affect the search performance.

In the evaluation tasks, we assume that *k*-*NN* graph Γ_k, β, a set of queries X_q, and a set of initial vertices X_0 are given, where X_0 is selected from the vertices in Γ_k uniformly at random independent of $q \in X_q$. For evaluation purposes only, we also assume that we know the correct solutions. Namely, we pre-calculate the set of the most similar objects $R(q)$ to given query q in each search evaluation task, although $R(q)$ is unknown in the actual search tasks.

Let $S(\Gamma_k,\beta)$ be a success ratio, which is the ratio of successfully finding $x^* \in R(q)$ in a range of given β. Then $S(\Gamma_k,\beta)$ is expressed by

$$S(\Gamma_k,\beta)=\frac{1}{|X_q||X_0|}\sum_{q\in X_q}\sum_{x_0\in X_0}\delta(q,x_0;\Gamma_k,\beta)\,, \tag{1}$$

where

$$\delta(q,x_0;\Gamma_k,\beta)=\begin{cases}1 & \text{if } BS(q,x_0,\Gamma_k,\beta)\in R(q)\\ 0 & \text{if } BS(q,x_0,\Gamma_k,\beta)\notin R(q).\end{cases} \tag{2}$$

The success ratio depends on two parameters: k and β.

We also introduce expected search cost $C(\Gamma_k)$ by eliminating β and depending only on a graph structure. $C(\Gamma_k)$ is expressed as follows:

$$C(\Gamma_k)=\frac{1}{|X_q||X_0|}\sum_{q\in X_q}\sum_{x_0\in X_0}\min\{\beta:\delta(q,x_0;\Gamma_k,\beta)=1\}\,. \tag{3}$$

$C(\Gamma_k)$ is the expected number of similarity calculations until finding $x^* \in R(q)$ in Γ_k by the *BS* algorithm with minimal upper bound β in each search task. Equivalently, $C(\Gamma_k)$ is the average of $|A|$ obtained when the *BS* algorithm with $\beta = |X|$ finds $x^* \in R(q)$, and is expressed as follows:

$$C(\Gamma_k)=\frac{1}{|X_q||X_0|}\sum_{q\in X_q}\sum_{x_0\in X_0}|A|\,. \tag{4}$$

For this evaluation task, we modify line 2 in **Algorithm** 2 to "$(|A|<\beta)\vee(z\notin R(q))$" so that the algorithm terminates immediately when it finds $z \in R(q)$. Note that $S(\Gamma_k,\beta)$ and $C(\Gamma_k)$ also depend on the random choice of X_0 and X_q, but we can omit its influence by setting $|X_q||X_0|$ sufficiently large.

4 *k*-Degree Reduced Nearest Neighbor Graph

Hereafter, we refer to the simple *k-NN* graph described in the previous section as Γ_{kNN} unless otherwise stated. Based on the simple *k-NN* graph Γ_{kNN}, we construct a more efficient nearest neighborhood graph in terms of the search costs associated with the *BS* algorithm. Suppose that, for query q, a neighborhood search from initial vertex x_0 to target vertex $x^* \in R(q)$ on any graph Γ is performed along the shortest path between them. In this optimal case, the search cost, which is the total number of similarity calculations, almost equals the total number of neighbors of the vertices along the path. The expected search cost is roughly estimated as the product of average degree $D(\Gamma)$ and average shortest path length $L(\Gamma)$ in Γ, i.e., $D(\Gamma)L(\Gamma)$. Therefore, we can improve the search performances if we can construct graph Γ whose average degree $D(\Gamma)$ is reduced over $D(\Gamma_{kNN})$, while keeping the increase in the average shortest path length $L(\Gamma)$ as small as possible.

We propose the *k-degree reduced nearest neighbor* (*k-DRNN*) graph Γ_{kDR} and the *degree reduction* (*DR*) algorithm that generates Γ_{kDR}. The basic idea of the *DR* algorithm is to construct graph Γ_{kDR} that has a much smaller average degree than the *k-NN* graph Γ_{kNN}, and has the following property: for each $x \in X$ and $y \in N_k(x)$, x and y may not be directly connected, but still x is *greedily reachable* from y; i.e., x is reachable from y by successive locally best moves in Γ_{kDR}. Its formal definition is given later. The *DR* algorithm guarantees that if the *k-NN* graph is "searchable" as a network, then the corresponding greedily reachable *k-DRNN* graph is also "searchable," and that if the *k-NN* graph is one connected component, so is the *k-DRNN* graph which is a subgraph of the *k-NN* graph. When Γ_{kDR} has the above property, we call that Γ_{kDR} is a *greedily reachable* network.

Algorithm 3. *DR Algorithm*: $DR(k, X, \Gamma_{(k-1)DR})$

Input: $k, X, \Gamma_{(k-1)DR}$
Output: Γ_{kDR}
1. Set $\Gamma_{kDR} \leftarrow \Gamma_{(k-1)DR}$
2. **for all** $x \in X$ **do**
3. Set $\{y\} \leftarrow N_k(x) \setminus N_{k-1}(x)$
4. **if** $GR(x, y, \Gamma_{kDR})$ = **FALSE then**
5. // add an edge (x, y)
   Set $V_{kDR}(x) \leftarrow V_{kDR}(x) \cup \{y\}$ and $V_{kDR}(y) \leftarrow V_{kDR}(y) \cup \{x\}$
6. **end if**
7. **end for**
8. Set $\Gamma_{kDR} \leftarrow (X, \cup_{x \in X} E_{kDR}(x))$, where $E_{kDR}(x) = \{(x, y) : y \in V_{kDR}(x)\}$

The *k-DRNN* graph is constructed recursively. The 1-*DRNN* graph Γ_{1DR} is the 1-*NN* graph Γ_{1NN} itself. For $k \geq 2$, *k-DRNN* graph is constructed from $(k-1)$-*DRNN* graph by the *DR* algorithm shown in Algorithm 3, where $V_{kDR}(x)$ is a set of adjacent vertices of x directly connected with the edges in Γ_{kDR}. At lines 4–6 in Algorithm 3, if x is not greedily reachable from y as judged by the *greedy-reachability checking* (*GR*) algorithm described in Algorithm 4, then x and y are mutually added to the set of adjacent vertices of each other.

Algorithm 4. *GR Algorithm*: $GR(x, y, \Gamma_{kDR})$

Input: $x \in X, y \in X, \Gamma_{kDR}$
Output: **TRUE** or **FALSE**

Set $z \leftarrow \arg\max_{z' \in V_{kDR}(y)} \{\rho(z', x)\}$
if $\rho(z, x) \geq \rho(y, x)$ **then**
   return TRUE
else
   return FALSE
end if

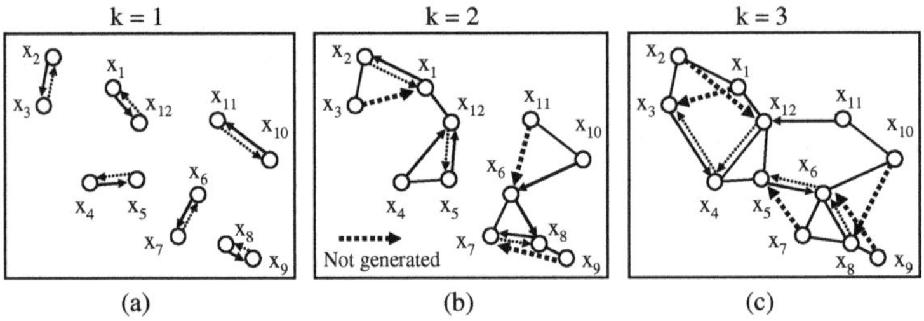

Fig. 2 The *DR* algorithm applied to 3-*DRNN* graph Γ_{3DR}

We prove that Γ_{kDR} is a greedily reachable network by recursion as follows. Obviously $\Gamma_{1DR} = \Gamma_{1NN}$ is a greedily reachable network. The *GR* algorithm evaluates whether there exists $z \in V_{kDR}(y)$ that is closer to x than y, and then returns **TRUE** if this is the case and **FALSE** otherwise. We only have to consider the case when such z exists for $y \in N_k(x)$. In this case, it follows from the definition of $N_k(x)$ that z belongs to $N_{k'}(x)$ for some $k' < k$. Thus, if $\Gamma_{k'DR}$ is a greedily reachable network, x is greedily reachable from z, and so is from y because of $z \in V_{kDR}(y)$. These above recursively prove that Γ_{kDR} is a greedily reachable network.

Figure 2 illustrates how the *DR* algorithm constructs 3-*DRNN* graph Γ_{3DR} consisting of 12 objects $\{x_1, x_2, \cdots, x_{12}\}$ (denoted as ○). Figure 2(a) shows the 1-*DRNN* graph, or equivalently, 1-*NN* graph generated by **Algorithm 1**. The originally directed edges from x_i $(i=1,2,\cdots,12)$ to $V_1(x_i)$ are changed to the undirected ones by line 2 in **Algorithm 1**. The 2-*DRNN* graph is shown in Fig. 2(b), where the edges in Γ_{1DR} and the newly generated edges in Γ_{2DR} are depicted with lines and arrows in the same way as the case of $k=1$, respectively. The edge denoted with a thick dashed arrow, for instance, from x_3 to x_1, is not generated because a greedily reachable path exists between the vertices. The final result (3-*DRNN* graph) is shown in Fig. 2(c). Three new edges were generated between the following object pairs: (x_3, x_4), (x_5, x_6), and (x_{11}, x_{12}).

5 Experiments

5.1 Object Sets

We extracted 1,298,933 documents from "*The New York Times*" from 1994 to 2006 for our experiments and prepared four data sets with different sizes that were selected uniformly at random: all documents (1-M-sized data set), 129,894 documents (100-K-sized data set), 12,990 documents (10-K-sized data set), and 1,299 documents (1-K-sized data set). Each data set was divided into two subsets including around 90% and 10% documents of the original set. The larger subsets were used as sets of the search objects and the smaller as sets of the queries. We counted the frequency of the occurrences of distinct words after stemming and stop word removal and then extracted features as real-value feature vectors using tf-idf, where

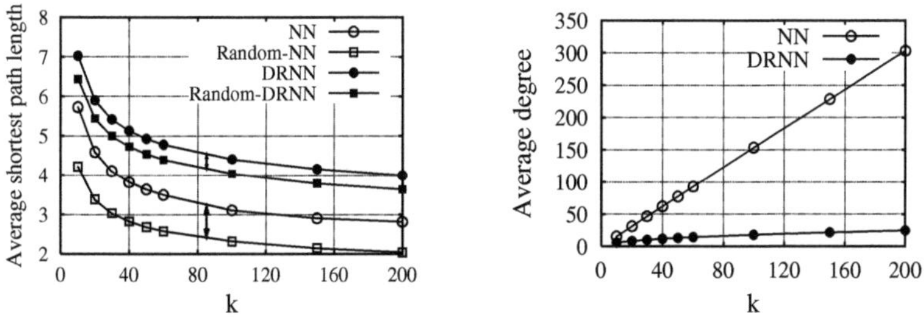

(a) Average shortest path length (b) Average degree

Fig. 3 Graph structures: (a) average shortest path length; (b) average degree against *k*

each dimension corresponds to each word. We adopted normalized cosine similarity as the similarity function between object pairs. The dimensionality (vocabulary size) increased with increases in the number of documents (document data size), which is known as Heaps' Law. Even in the case of the smallest 1-K-sized data set, the dimensionality reached 26,949, resulting in a similarity search problem in a high-dimensional metric space.

5.2 Graph Structures

We analyze the structures of the *k-NN* and the *k-DRNN* graphs in this section. In particular, we focus on the two structural properties: the average shortest path length $L(\Gamma)$ and the average degree $D(\Gamma)$.

As an example, we show the properties of the graphs constructed from the 100-K-sized data set in Fig. 3, where the *k-NN* graph with k=8, thus *k-DRNN* graph with k=8 as well, is confirmed to have one connected component. Figure 3(a) shows the average shortest path length $L(\Gamma)$ and the average degree $D(\Gamma)$ of the *k-NN* and the *k-DRNN* graphs with those of the corresponding random graphs, Random-NN and Random-DRNN, which are randomly generated so that their average degrees equal those of the corresponding *k-NN* and *k-DRNN* graphs, respectively.

We can observe from Fig. 3(a) that the average shortest path lengths of the *k-NN* and the *k-DRNN* graphs, in which similar and closely related objects are directly connected, are almost as small as those of the corresponding randomly connected graphs. These results strongly suggest that both the *k-NN* and *k-DRNN* graphs are small-world networks and "searchable", as the acquaintance network in Milgram's experiment.

From Fig. 3(a), we can also observe that the ratio of the average shortest path length of the *k-DRNN* graph to that of the *k-NN* graph is small and at most 1.4. Moreover, Fig. 3(b) shows that the average degree of the *k-DRNN* graph and its increase with *k* are much smaller than those of the *k-NN* graph. From these properties of the average degree $D(\Gamma)$ and the average shortest path length $L(\Gamma)$ of the *k-DRNN* and the *k-NN* graphs, we can expect that the *k-DRNN* graph is more suitable than the *k-NN* graph as an index structure for graph-based similarity searches.

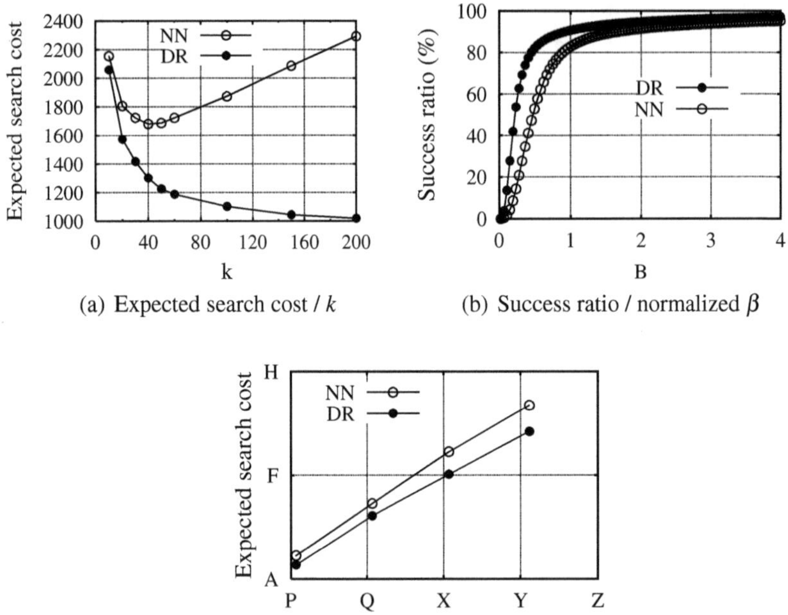

(a) Expected search cost / k

(b) Success ratio / normalized β

(c) Expected search cost / number of documents

Fig. 4 Search performances of *NN* and *DR* search methods: (a) expected search cost against k; (b) success ratio against normalized β; (c) expected search cost against number of documents

5.3 Search Performance Evaluation

We abbreviate the best-first neighborhood search method with the simple *k-NN* graph to the *NN* search method and that with the *k-DRNN* graph to the *DR* search method. We experimentally evaluate the expected search cost $C(\Gamma)$ and success ratio $S(\Gamma,\beta)$ of the *NN* and the *DR* search methods using the document data sets with different sizes. In the experiments, we set 10 as the number of initial nodes $|X_0|$ for each query $q \in X_q$ in Eqs. (1) and (4). As an example of the evaluation results, we first show the expected search cost and the success ratio obtained by applying the *NN* and *DR* methods to the 100-K-sized data set ($|X|$=116,905) in Figs. 4(a) and 4(b), respectively.

As shown in Fig. 4(a), $C(\Gamma_{kNN})$ obtained by the *NN* method has a minimum value of 1,681 when k=40, namely, $C(\Gamma_{40NN})$=1,681 corresponding to 1.44% of $|X|$. In contrast, $C(\Gamma_{kDR})$ obtained by the *DR* method decreases monotonically in the range of $8 \leq k \leq 200$. $C(\Gamma_{100DR})$=1,102 (0.94% of $|X|$) and $C(\Gamma_{200DR})$=1,019 (0.87% of $|X|$). Both search methods found the target vertex to the given query with extremely low expected search cost, compared with a brute-force search method that requires an average search cost of 50% of $|X|$. Moreover, a comparison of the results obtained by the two methods reveals that the *DR* method has the following two advantages over the *NN* method. The first is lower expected search cost, and the second is lower sensitivity of the expected search cost to k in the relatively broader range, which enables us to determine the value of k without difficulty.

Figure 4(b) shows the success ratio against β normalized by $|X|$. The success ratios were obtained by using 40-*NN* graph Γ_{40NN} and 100-*DRNN* graph Γ_{100DR}, where k=40 and k=100 were chosen so that the expected search cost with the *NN* and the *DR* methods were the best, respectively. The results show experimentally that both methods successfully found the exact solutions with very high probability under an extremely low value of normalized β. For instance, the methods found a target vertex to a given query with a 90% success ratio, if the values of $\beta/|X|$ for the *NN* and the *DR* methods were set at 1.70% and 0.94%, respectively.

Next, we focus on the relationship between the minimum expected search cost in the range of $k \leq 200$ and the size of the document data sets, as shown in Fig. 4(c). The result shows that the relationship can be approximated by sublinear functions for both methods. The exponents of the functions of the *DR* and the *NN* methods are 0.42 and 0.48, respectively. Judging from this fact, the methods would be applicable to very large-scale data sets.

6 Conclusion

We presented a novel graph-based approach for fast similarity searches suitable for large-scale and high-dimensional data. We focused on the well-known feature that small-world networks are "searchable" and proposed an efficient index structure called the degree-reduced nearest neighbor graph based on similarity defined between objects. A similarity search was then formulated as a problem of finding the most similar object to a query object by following the links in this graph with a best-first neighborhood search (*BS*) algorithm. Experimental results showed that the proposed search method achieved a high success ratio with low expected search cost. In particular, it found a nearest neighbor object from nearly one million documents with average similarity evaluations around only 0.9% of all the documents.

We employed the *BS* algorithm among many candidate search strategies applicable to the graph-based search algorithm to confirm that our "searchable" small-world network approach is, in fact, effective for similarity search. The *BS* is suitable in centralized environments because it exploits global shared memory to store already calculated vertices. In more decentralized or distributed environments, a depth-first search *DS* or a greedy search *GS* may be more suitable than the *BS*.

Since the proposed method used the upper bound β of search costs as a termination criterion, it could not always find the exact solution in the range of β. Instead, the upper bound of similarity between the query and the solution may be used as another termination criterion. However, to guarantee that the algorithm terminates in a finite amount of time, we need to determine the distribution of the objects in advance, which is difficult in practice.

The advantage of our proposed method is that it can considerably reduce the number of similarity evaluations and is applicable to any type of data including symbol sequences, images, and speech data sets with any type of similarities (or distances) including those that are highly time consuming to compute.

Acknowledgements. The authors would like to thank H. Fujimoto for his support on experiments. This work was supported by KAKENHI (20500109).

References

1. Watts, D.J., Strogatz, S.H.: Collective dynamics of 'small-world' networks. Nature 393, 440–442 (1998)
2. Kleinberg, J.: Complex networks and decentralized search algorithms. In: Proc. Int. Congress of Mathematicians (2006)
3. Watts, D.J., Dodds, P.S., Newman, M.E.J.: Identity and search in social networks. Science 296, 1302–1305 (2002)
4. Milgram, S.: The small world problem. Psychology Today 2, 60–67 (1967)
5. Chávez, E., Navarro, G., Baeza-Yates, R., Marroquín, J.L.: Searching in metric spaces. ACM Comp. Surveys 33, 273–321 (2001)
6. Kleinberg, J.: The small-world phenomenon: an algorithmic perspective. In: Proc. ACM Symp. Theory of Computing, pp. 163–170 (2000)
7. Orchard, M.T.: A fast nearest-neighbor search algorithm. Proc. Int. Conf. Acoust., Speech, Signal Process. 4, 2297–2300 (1992)
8. Sebastian, T.B., Kimia, B.B.: Metric-based shape retrieval in large databases. In: Proc. Int. Conf. Pattern Recognition, vol. 3, pp. 291–296 (2002)
9. Adamic, L.A., Lukose, R.M., Puniyani, A.R., Huberman, B.A.: Search in power-law networks. Phys. Rev. E 64, 046135 (2001)
10. Androutsos, P., Androutsos, D., Venetsanopoulos, A.N.: Small world distributed access of multimedia data: An indexing system that mimics social acquaintance networks. IEEE Signal Processing Magazine 23, 142–153 (2006)
11. Lin, C.-J., Tsai, S.-C., Chang, Y.-T., Chou, C.-F.: Enabling keyword search and similarity search in small-world-based P2P systems. In: Proc. 16th Int. Conf. on Computer Communications and Networks, pp. 115–120 (2007)
12. Andoni, A., Datar, M., Immorlica, N., Indyk, P., Mirrokni, V.: Locality-sensitive hashing using stable distributions. In: Nearest-neighbor methods in learning and vision. MIT Press, Cambridge (2005)
13. Bustos, B., Navarro, G., Chávez, E.: Pivot selection techniques for proximity searching in metric spaces. Pattern Recog. Lett. 24, 2357–2366 (2003)
14. Şimşek, Ö., Jensen, D.: Decentralized search in networks using homophily and degree disparity. In: Proc. 19th Int. Joint Conf. on Artificial Intelligence, pp. 304–310 (2005)

Detection of Packet Traffic Anomalous Behaviour via Information Entropy

Anna T. Lawniczak, Hao Wu, and Bruno N. Di Stefano

Abstract. Spatio-temporal dynamics of packet traffic in data networks is complex and its monitoring is a challenging task. We study if information entropy of packet traffic monitored at selected set of nodes may provide a method for monitoring network-wide behaviour of packet traffic and for detection of anomalous traffic, e.g., distributed denial-of-service attacks. We conduct our investigation for a packet switching network model for static and dynamic routings. We show that the proposed information entropy method may detect changes in "natural" randomness of spatio-temporal distributions of packets among routers caused by anomalous traffic and that the emerging anomalies are easier to detect for DDoS attacks with larger number of attackers and/or on networks using static rather than dynamic routing.

Keywords: packet switching network, distributed denial of service attack, detection, entropy.

1 Introduction

The Internet is one of the best known examples of packet switching network (PSN) and it is one of the most complex ones. Its various subnets and many other PSNs are constantly under different types of network-wide attacks that may drastically slow down their operations and response to legitimate network activities.

Anna T. Lawniczak and Hao Wu
Department of Mathematics and Statistics, University of Guelph
Guelph, Ontario N1G 2W1, Canada
The Fields Institute For Research in Mathematical Science
Toronto, Ontario M5T 3J1, Canada
e-mail: alawnicz@uoguelph.ca, kamanwu@gmail.com

Bruno N. Di Stefano
Nuptek Systems Ltd., Toronto, Ontario M5R 3M6, Canada
e-mail: b.distefano@ieee.org

R. Menezes et al. (Eds.): Complex Networks: CompleNet 2009, SCI 207, pp. 197–208.
springerlink.com © Springer-Verlag Berlin Heidelberg 2009

The inherent network complexity due to the coupling of topology, routing algorithm, incoming traffic, and a multitude of asynchronously communicating independent entities, is made even more complex by the presence of attacks of unknown modality originating at a large number of unknown sites and targeting an unknown number of victim sites. To detect anomalous traffic packets headers, packets aggregate flows, and/or correlations are analysed with the aim of distinguishing normal traffic from attack traffic [1]. This task is of a formidable challenge, in particular for network-wide attacks. We investigated the use of information entropy of packet traffic monitored at selected sets of routers to detect anomalous packet traffic. Our focus is on the distributed denial-of-service (DDoS) attack directing a huge number of "ping" requests to the target victim of the attack [2, 3]. We conduct this study via a PSN model.

Being purposely created by humans, DDoS attacks must affect the natural "randomness" and "natural structure and order" of packet traffic under normal conditions [4]. Thus, in turn they must affect the "natural" information entropy profiles of "normal" packet traffic and cause shifts in these profiles, [3, 4]. We verify this by simulation. The entropy profiles of packet traffic monitored at even small sets of nodes may change significantly almost immediately after the DDoS attacks start. Thus, they detect these attacks and are useful tool in detection of anomalous packet traffic. We notice that it is easier to detect DDoS attacks when using static rather than dynamic routing. Also, our simulations show that the entropy profiles of packet traffic monitored at selected nodes may convey information about the network-wide packet traffic behaviour.

We describe our PSN model [5, 6], its C++ simulator (Netzwerk) [7, 8], and its customization to model and simulate a "ping" type DDoS attacks. We define entropy functions of packet traffic and use them to detect of DDoS attacks in our simulation experiments. We present selected simulation results and our conclusions.

2 Descriptions of PSN and DDoS Attacks Models

To study DDoS attacks we customized our PSN model described in details in [5, 6]. This model is an abstraction of the Network Layer of the 7-Layer OSI Reference Model [9]. Our PSN model is concerned primarily with packets and their routings; it is scalable, distributed in space, and time discrete. It avoids the overhead of protocol details present in many PSN simulators designed with different aims in mind than study of macroscopic network-wide dynamics of packet traffic. A PSN connection topology is viewed as a weighted directed multigraph $\boldsymbol{L}$ where each node/router corresponds to a vertex and each communication link is represented by a pair of parallel edges oriented in opposite directions. In each PSN model setup each cost of transmission of a packet along a link (an edge) is computed using the same type of *edge cost function* (*ecf*) that is either the *ecf* called *ONE* (ONE), or *QueueSize* (QS), or *QueueSizePlusOne* (QSPO). The *ecf* ONE assigns a value of "one" to all edges in the lattice $\boldsymbol{L}$. This results in a *static routing* since this value does not change during the course of a simulation. The *ecf* QS assigns to each edge in the lattice $\boldsymbol{L}$ a value equal to the length of the outgoing queue at the node from which the edge originates. The *ecf* QSPO assigns a value that is the sum of a constant "one" plus the length of the outgoing queue at the node from which the edge originates. The routing decisions made using *ecf* QS or QSPO result in *adaptive* or *dynamic routing* because they rely on the current state of the network simulation and the packets are routed avoiding congested nodes during the PSN model simulation. In our PSN model, each packet is transmitted via routers from its source to its destination according to the routing decisions made independently at each router and based on a *minimum least-cost criterion* of selecting a *shortest path* from a packet current node to its destination. Thus, if the PSN model is setup with *ecf* ONE then the routing is the *minimum hop routing* (*minimum route distance*) and if it is setup with *ecf* QS or QSPO then it is the *minimum length routing*. It is important to notice, that in the case of PSN model setup with *ecf* QS or QSPO, because these costs are dynamic, each packet is forwarded from its current node to the next one that belongs to a least cost shortest path from the packet current node to its destination at this time. The PSN model uses *full-table routing*, that is, each node maintains a routing table of least path cost estimates from itself to every other node in the network. The routing tables are updated at each time step when the *ecf* QS or QSPO is used; see [5, 6]. Since the values of the *ecf* ONE do not change over time the routing tables do not need to be updated for the static *ecf* ONE; see [5, 6]. We update the routing tables using distributed routing table update algorithm [6].

In our simulations to study DDoS attacks we use a version of PSN model in which each node performs the functions of *host* and *router* and maintains one incoming and one outgoing queue which is of unlimited length and operates according to a first-in, first-out policy, see [6] for other options. At each node, independently of the other nodes, packets are created randomly with probability λ called *source load*. In our PSN model all messages are restricted to one packet carrying time of creation, destination address, and number of hops taken.

In the PSN model time is discrete and we observe its state at the discrete times $k = 0, 1, 2, \ldots,$ T, where T is the final simulation time. In the presented simulations each PSN model setup is characterize by the selection of network connection topology, *ecf* type and source load value. At time $k = 0$, the setup of the PSN model is initialized with empty queues and the routing tables are computed. The time discrete, synchronous and spatially distributed PSN model algorithm consists of the sequence of five operations advancing the simulation time from k to $k + 1$. These operations are: (1) Update routing tables, (2) Create and route packets, (3) Process incoming queue, (4) Evaluate network state, (5) Update simulation time. The detailed description of this algorithm is provided in [5, 6].

We modified the described PSN model to study DDoS attacks by selecting one victim computer and a user defined number of zombies either located at specified nodes or located at random. For each zombie start and end of attack time can be specified separately. As in most real life cases, zombies continue to carry on their normal jobs during an attack, i.e. they act also as sources, destinations, and routers of legitimate data transfers. However, each zombie also sends a packet to the victim at each time step of an attack simulation.

3 Entropy Functions of Packet Traffic

To monitor packet traffic behaviour we use the *entropy functions of packet traffic* monitored at a set of nodes of PSN model introduced in [4]. These functions are defined as follows. Let M be a set of N monitored nodes. The set M may include all network nodes except zombies and the victim. We index all nodes in the set M by the parameter i (i.e., $i = 1, \ldots, N$). We denote by $q(i,k)$ a number of packets at the outgoing queue of a node i at time k. At each time k we calculate probability density function $p(i,k)$ of packets queuing at a node i of the set M as follows

$$p(i,k) = q(i,k) / \sum_{i=1}^{N} q(i,k).$$

We calculate *entropy function of packet traffic* monitored at nodes of the set M as

$$H(M,k) = -\sum_{i=1}^{N} p(i,k) \log p(i,k),$$

using convention that if $p(i,k) = 0$, then $p(i,k)\log p(i,k) = 0$.

For each PSN model setup operating under normal conditions, i.e. in the absence of any attack, and for each considered set M of monitored nodes we calculate first entropy function $H(M,k)$. This “natural” entropy function is a sort of “fingerprint” profile of “normal” packet traffic passing through the nodes of the set M when PSN model operates under normal conditions. Next, for each considered PSN model setup and for each considered set M of monitored nodes we calculate the entropy function $H(M,k)$ when network is under a DDoS attack. In our study we consider sets M of different sizes (i.e., having different numbers of monitored nodes) and for each set size we consider sets M that differ only in locations of the monitored nodes. We select locations of monitored nodes randomly using different seeds of random number generator and exclude the location of the victim and the zombies. We make this exclusion because we are interested in the dynamics of packet traffic passing through the “normal” nodes, i.e. which are not the victim or the zombies.

4 DDoS Attacks Simulation Scenarios

We discuss selected simulation results of DDoS attacks on PSN model with $L^p_\square(37, ecf, \lambda)$ setups, i.e. on PSN models with network connection topology isomorphic to $L^p_\square(37)$ (i.e., periodic square lattice with 37 nodes in the horizontal and vertical directions) and for each of the ecf = ONE, or QS, or QSPO, and *source load* value $\lambda = 0.040$. At this value packet traffic of each PSN model setup is in free flow state, i.e. is free of any congestion. The number of packets in transit does not increase with time and fluctuates around some constant value after an initial transient time (recall, we start each simulation of a PSN model setup always with empty queues). This can be seen from Figure 1 showing queue sizes of the nodes of the PSN model with $L^{P}_{\square}(37, \text{QSPO}, 0.040)$ setup. The phase transition point from free flow to congested network state, i.e. the *critical source load* value λ_c for each of the considered PSN model setups is as follows: $\lambda_c = 0.053$ for $L^{P}_{\square}(37, \text{ONE})$, and $\lambda_c = 0.054$ for $L^{P}_{\square}(37, \text{QS})$ and $L^{P}_{\square}(37, \text{QSPO})$.

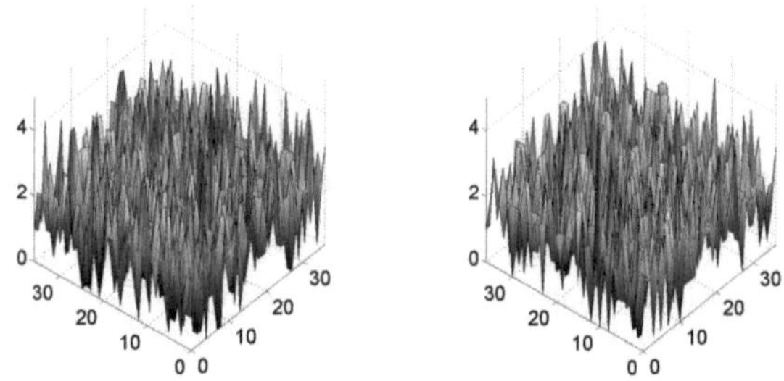

Fig. 1 The plots of queue sizes at the nodes of the PSN model with $L^{P}_{\square}(37, \text{QSPO}, 0.040)$ setup at time $k=2^{13}$ (i.e., at k=8,192, left plot) and $k=2^{14}$ (i.e., at k=16,384 right plot)

All DDoS attacks we start after the initial transient time, i.e. when the network operates already in its normal steady state for some time. For the presented simulation results we selected time $k_0 = 20{,}480$ as start of all DDoS attacks. This time is much longer than the transient times. All the DDoS attacks lasted until the final simulation time, $T = 131{,}072$, being the same for all PSN model setups. In our simulations we considered a series of separate DDoS attacks on a victim having always the same location in all the experiments. Each attack is characterized by a number of active attackers/zombies. In this series of attacks, while increasing number of zombies we maintained always the same locations of the zombies from the DDoS attacks with their lower numbers, i.e. each time we only added new zombies to the set of the zombies from the previous attack. In this paper we present the results for the DDoS attacks with number of active attackers/zombies varying from 0 to 6, i.e. with the number of zombies varying from 0% to about 0.44% of the total number of nodes in the network.

5 Discussions of DDoS Attack Simulation Results

We explore usefulness of entropy functions of packet traffic monitored at a set of selected nodes for early detection of anomalous packet traffic. In our study we consider static and adaptive routing.

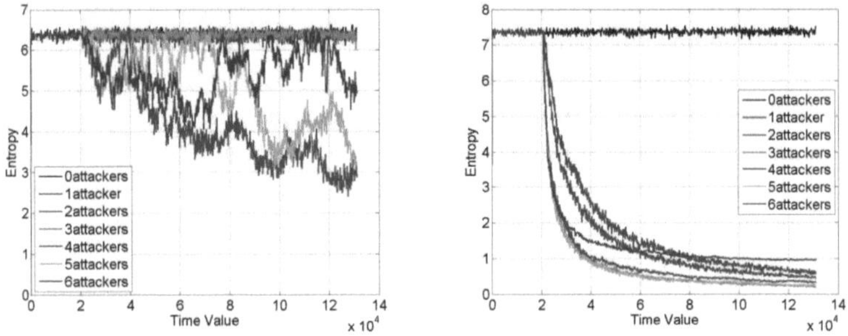

Fig. 2 Time dependent graphs of "natural entropy" profiles (i.e., with 0 attackers) and entropy functions of packet traffic monitored at 10% (left plot) and 20% (right plot) of normal nodes selected randomly with seed 1 during DDoS attacks in PSN model with $L^{P}_{\square}(37, \text{ONE}, 0.040)$ setup. The colours of the graphs correspond to the number of attackers varying from 0 to 6 and they are explained in the plot legends. Each DDoS attack starts at k_0=20480. In the left plot the graphs of entropy functions corresponding to the attacks with 1, 2 and 3 attackers are superimposed with the graph of the "natural entropy" profile

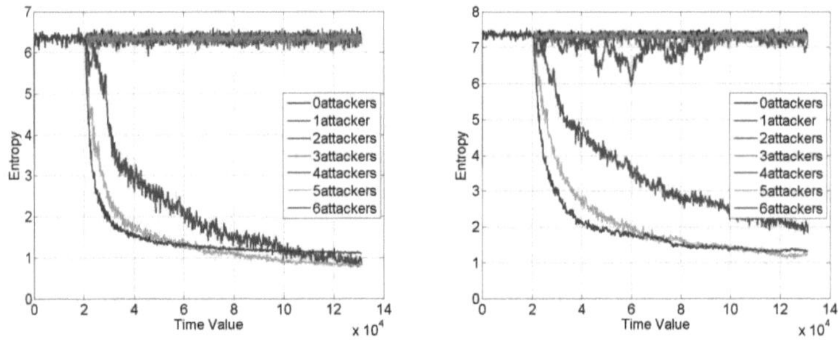

Fig. 3 Time dependent graphs of "natural entropy" profiles (i.e., with 0 attackers) and entropy functions of packet traffic monitored at 10% (left plot) and 20% (right plot) of normal nodes selected randomly with seed 2 during DDoS attacks in PSN model with $L^{P}_{\square}(37, \text{ONE}, 0.040)$ setup. The colours of the graphs correspond to the number of attackers varying from 0 to 6 and they are explained in the plot legends. Each DDoS attack starts at k_0=20480. The graphs of entropy functions corresponding to the attacks with 1, 2 and 3 attackers are superimposed with the graph of the "natural entropy" profile in the left plot and they are partially superimposed in the right plot. The graphs of entropy functions corresponding to the attacks with 4, 5, and 6 attacks sharply decrease almost immediately after the start of each of these DDoS attacks

Figure 2 to Figure 5 display time dependent graphs of "natural entropy" profiles (i.e., with 0 attackers) and the time dependent graphs of entropy functions of packet traffic monitored at 10% (left plots in Figure 2 to Figure 5) and 20% (right plots in Figure 2 to Figure 5) of all "normal" nodes selected randomly with seed 1 (Figure 2 and Figure 4) and seed 2 (Figure 3 and Figure 5) during DDoS attacks in PSN model with $\boldsymbol{L}^{P}_{\square}(37, \text{ONE}, 0.040)$ setup (Figure 2 and Figure 3) and with

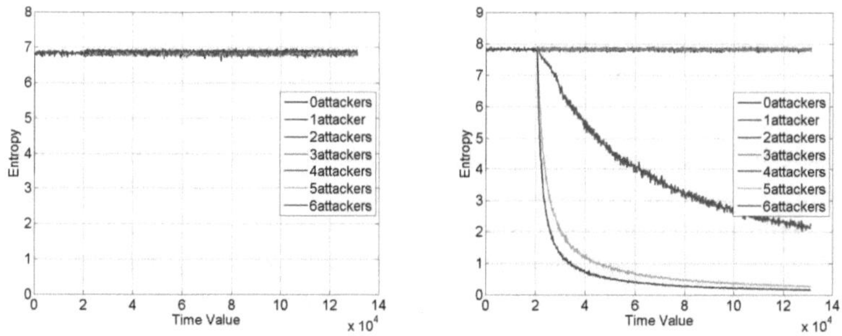

Fig. 4 Time dependent graphs of "natural entropy" profiles (i.e., with 0 attackers) and entropy functions of packet traffic monitored at 10% (left plot) and 20% (right plot) of normal nodes selected randomly with seed 1 during DDoS attacks in PSN model with $L^{P}_{\square}$(37, QSPO, 0.040) setup. The colours of the plots correspond to the number of attackers varying from 0 to 6 and they are explained in the plot legends. Each DDoS attack starts at k_0=20480. In the left plot all the graphs of entropy functions corresponding to the DDoS attacks are superimposed. In the right plot the graphs of entropy functions corresponding to the DDoS attacks with 1, 2 and 3 attackers are superimposed with the graph of the "natural entropy" profiles

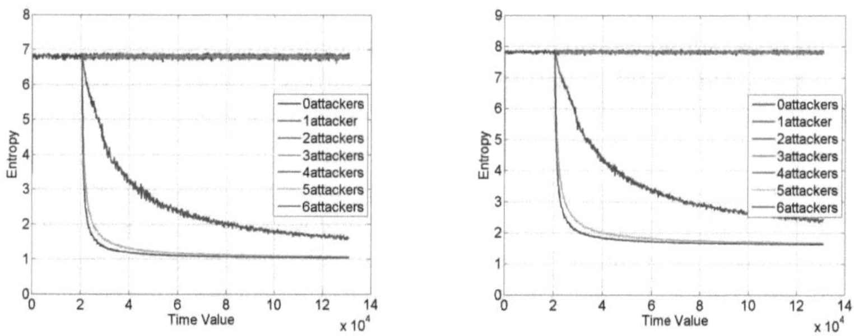

Fig. 5 Time dependent graphs of "natural entropy" profiles (i.e., with 0 attackers) and entropy functions of packet traffic monitored at 10% (left plot) and 20% (right plot) of normal nodes selected randomly with seed 2 during DDoS attacks in PSN model with $L^{P}_{\square}$(37, QSPO, 0.040) setup. The colours of the graphs correspond to the number of attackers varying from 0 to 6 and they are explained in the plots legends. Each DDoS attack starts at k_0=20480. The graphs of entropy functions corresponding to the DDoS attacks with 1, 2 and 3 attackers are superimposed with the graph of the "natural entropy" profiles

$L^{P}_{\square}$(37, QSPO, 0.040) setup (Figure 4 and Figure 5). In these figures the colour coding of the graphs of the entropy functions corresponds to the number of attackers from 0 to 6, it is explained in the plots legends and is consistent across the figures.

Looking at the plots of Figure 2 to Figure 5 we observe that the "natural entropy" profiles fluctuate slightly around constant valued functions and that these constant values increase with the increase of the number of monitored nodes for both PSN model setups. The increase of these constant values of "natural" entropy functions is due the increased randomness in queue size distributions among the monitored nodes when their number increases.

In the left plots of Figure 2 and Figure 3 we observe that the graphs of entropy functions of packet traffic monitored at 10% of all normal nodes (i.e., at 136 out of 1369 nodes selected randomly, respectively, with seed 1 or seed 2) almost immediately deviate significantly downward from the "natural" entropy profile for each DDoS attack with number of attackers 4, or 5, or 6. This means that the entropy functions of packet traffic monitored at the considered sets of nodes detect an anomalous packet traffic caused by the DDoS attacks in the PSN model with $\boldsymbol{L}^{P}_{\square}(37, \mathrm{ONE}, 0.040)$ setup almost immediately after the start of each of these attacks. Because, when the values of entropy functions sharply decrease from their "fingerprint" profiles this means that these functions detect with certainty the presence of an infrequent event, i.e. an emerging anomaly in packet traffic, and in our case "ping" type DDoS attack.

Looking at the left plot of Figure 5, we observe that the entropy functions of packet traffic monitored at 10% of all normal nodes selected randomly with seed 2 also, almost immediately detect an anomalous packet traffic caused by DDoS attacks with number of attackers 4, or 5, or 6 in the PSN model with $\boldsymbol{L}^{P}_{\square}(37, \mathrm{QSPO}, 0.040)$ setup. However, from the left plot of Figure 4 we see that these DDoS attacks in the PSN model with $\boldsymbol{L}^{P}_{\square}(37, \mathrm{QSPO}, 0.040)$ setup are not detected by the entropy functions when the set of 10% of all normal nodes selected randomly with seed 1 is used instead. In this case the graphs of entropy functions of the monitored packet traffic do not decrease and they are superimposed with the "natural" entropy profile of packet traffic.

Let us mention that the respective graphs of entropy functions of packet traffic monitored at 5% of all normal nodes selected randomly with seed 1, or seed 2 look qualitatively very similar to those displayed, respectively, in Figure 2 to Figure 5, see [4]. Also, when the PSN model uses *ecf* QS instead of *ecf* QSPO the respective graphs of entropy functions of monitored packet traffic are qualitatively very similar to those displayed in Figure 3 and Figure 5.

From Figure 2 to Figure 5 we see that the entropy functions of packet traffic monitored at the considered two sets of selected nodes did not detect the DDoS attacks with the number of attackers 1, or 2, or 3. When we increased the size of these two monitored sets to 20% of all normal nodes selected randomly (i.e., to 273 nodes out of 1,369 nodes) some of these attacks in the PSN model with $\boldsymbol{L}^{P}_{\square}(37, \mathrm{ONE}, 0.040)$ setup were detected, see the right plot of Figure 2 and of Figure 3. However, in the PSN model with $\boldsymbol{L}^{P}_{\square}(37, \mathrm{QSPO}, 0.040)$ setup they were still not detected, see the right plot of Figure 4 and of Figure 5. The reasons that for the PSN model with $\boldsymbol{L}^{P}_{\square}(37, \mathrm{QSPO}, 0.040)$ setup these attacks were not detected by the entropy functions of packet traffic, even when we used 100% of all normal nodes as the monitoring set (results not shown here) are as follows: the normal traffic of the PSN model is low, the adaptive routing balances rather evenly the distribution of packets among the network nodes and the capacity of

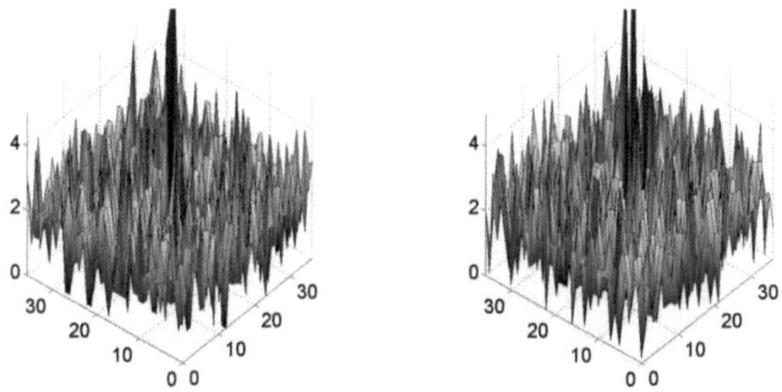

Fig. 6 Snapshot at time $k = 2^{15}$ (i.e., 32,768, left plot) and at time $k = 2^{16}$ (i.e., 65,536, right plot) of a spatial distribution of queue sizes in the PSN model with $\boldsymbol{L}^{P}_{\square}(37, \text{QSPO}, 0.040)$ setup being under DDoS attack with 3 attackers. The maximum queue size excluding the zombie queues is 9 at $k = 2^{15}$ and 12 at $k = 2^{16}$.

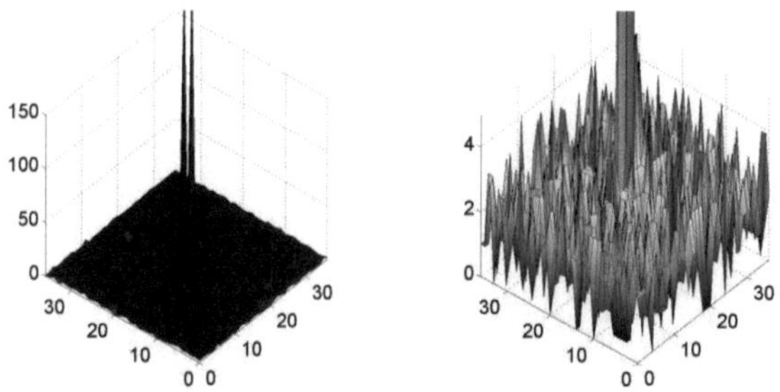

Fig. 7 Snapshot at time $k = 2^{15}$ (i.e., 32,768) of a spatial distribution of queue sizes in the PSN model with $\boldsymbol{L}^{P}_{\square}(37, \text{QSPO}, 0.040)$ setup being under DDoS attack with 4 attackers. The right plot is the magnification of the left one. The maximum queue size excluding the zombies queues is 257 at $k = 2^{15}$

the victim to accept the incoming packet traffic (i.e., 4 packets at each time step) is high enough to absorb the incoming packets generated by the DDoS attacks with 1, or, 2, or 3 attackers. Comparing Figure 6, i.e. for the case of the DDoS attack with 3 attackers, with Figure 1, i.e. for the case of normal packet traffic, we see that the randomness of the distribution of queue sizes is not significantly altered by this DDoS attack during the simulation and we see only slightly increase in the queue sizes of the nodes in direct neighbourhood of the victim. The same is true in the case of an attack with 1, or 2 attackers. Because the natural randomness of the distributions of packets among the nodes is not significantly altered by these weak DDoS attacks (i.e., with number of attackers 1, or 2, or 3) also the entropy functions of packet traffic are not significantly altered and their graphs are superimposed with the "natural" entropy profiles, see Figure 4 and Figure 5.

When the number of attackers in the PSN model with $\boldsymbol{L}^{P}_{\square}(37, \text{QSPO}, 0.040)$ setup is greater than 3 the amount of packet traffic generated by the zombies

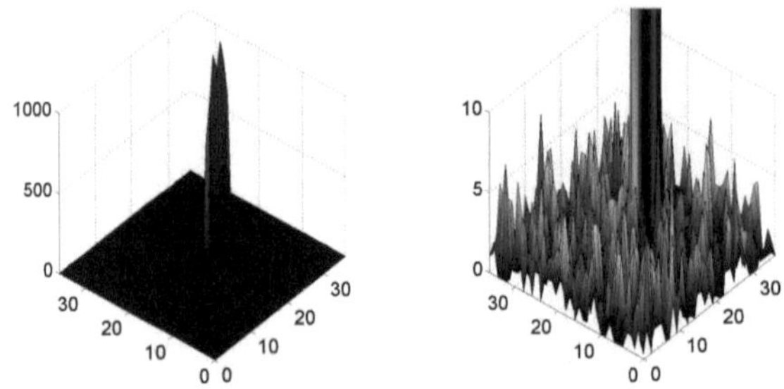

Fig. 8 Snapshot at time $k = 2^{15}$ (i.e., 32,768) of a spatial distribution of queue sizes in the PSN model with $\boldsymbol{L}^{P}_{\square}(37, \text{ONE}, 0.040)$ setup being under DDoS attack with 1 attacker. The right plot is the magnification of the left one. The maximum queue size excluding the zombies' queues is 1,306

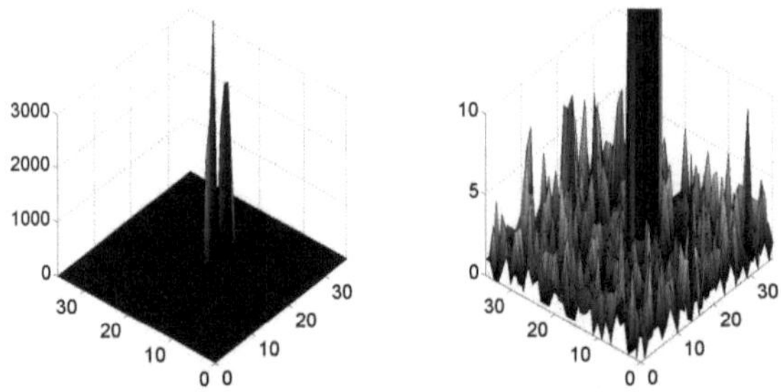

Fig. 9 Snapshot at time $k = 2^{15}$ (i.e., 32,768) of a spatial distribution of queue sizes in the PSN model with $\boldsymbol{L}^{P}_{\square}(37, \text{ONE}, 0.040)$ setup being under DDoS attack with 3 attackers. The right plot is the magnification of the left one. The maximum queue size excluding the zombies' queues is 4,415

outbalances the victim capacity to accept it and the local congestion builds up quickly around the victim, changing the natural randomness in the distribution of packet traffic. This can be seen from Figure 7. The qualitative shape of the distribution of queue sizes (i.e., the local congestion only around the victim) is maintained during the simulation for all the considered DDoS attacks with number of attacker greater than 3, as the total amount of the network incoming traffic does not exceed the critical source load value. Thus, even for some small sets of monitored nodes the values of entropy functions of packet traffic may decrease very rapidly almost immediately after the start of DDoS attacks, as can be seen from the right plot of Figure 4 and both plots of Figure 5. In these plots the graphs of entropy functions of packet traffic look qualitatively very similar to those obtained for the set of monitoring nodes consisting of 100% of all normal nodes. Thus, the entropy functions of packet traffic of properly selected small sets of monitoring nodes may detect the onset of DDoS attacks and may also provide and maintain the information about the behaviour of packet traffic in the whole network.

Our simulations show that these tasks are more easily achieved if the PSN model uses static routing (i.e., *ecf* ONE) instead of the dynamic one (i.e., *ecf* QS, or QSPO). The reason for this is that, when the PSN model uses the static routing even a weak DDoS attack (e.g., even with 1 attacker) causes a very quick build up of congestion along the shortest paths from the attacker or the attackers to the victim, as can be seen from Figure 8 (e.g., 1 attacker) and Figure 9 (e.g., 3 attackers). Because packets do not have an ability to avoid congested nodes on their routes from sources to destinations, the local congestions along the shortest paths from the zombies to the victim grow very quickly with time, even in the case of 1 attacker, while the queue sizes outside these shortest paths remain more or less of the same magnitude, see Figure 8 and Figure 9. Thus, even for a weak DDoS attacks in PSN model with $\boldsymbol{L}^{P}_{\square}(37, \text{ONE}, 0.040)$ setup, new type of order emerges in packet traffic that it is less uniform. This new type of emerged randomness in queue size distribution is sufficiently different from the natural one and the entropy functions of packet traffic are able to detect it for properly selected sets of monitoring routers, compare the right plot of Figure 2 with the right plot of Figure 3. On the right plot of Figure 2 we observe that all the graphs of entropy functions of the monitored packet traffics decrease very rapidly almost immediately after the start of each of the DDoS attacks, until they reach new levels significantly lower than the "natural" entropy profile. Additionally, the graphs of these entropy functions of packet traffics are qualitatively very similar to the ones calculated for 100% of all normal nodes (graphs not shown here). Thus, once again we observe that for properly selected even relatively small sets of monitoring routers the entropy functions of packet traffic may detect onset of even a weak DDoS attacks, or other types of anomalous packet traffic and they may also provide and maintain the information about the behaviour of packet traffic in the whole network.

6 Conclusions

We have demonstrated that the entropy function of packet traffic monitored at selected network nodes seems to be a promising tool in detection of anomalous packet traffic, e.g. DDoS attacks. This method may detect changes in "natural" randomness of spatio-temporal distributions of packets among nodes caused by anomalous packet traffic, e.g. DDoS attacks. This is because for properly selected sets of monitored nodes the values of entropy functions of packet traffic sharply decrease from the "fingerprint" profiles shortly after start of each DDoS attack, meaning with certainty a presence of an infrequent event. The emerging anomalies in packet traffic are easier to detect for DDoS attacks with larger number of attackers and/or on networks using static routing instead of a dynamic one. We have observed that to detect weak DDoS attacks one needs to monitor larger number of nodes and/or pay attention to their selection within the network. Some of the questions that need to be explored further are how to select the monitored nodes and how many of them so that entropy functions can provide a reliable method for monitoring network-wide behaviour of packet traffic.

Acknowledgments. The authors acknowledge the prior work of A.T. Lawniczak (A.T.L.) with A. Gerisch and the use of Sharcnet computational resources. A.T.L. acknowledges partial financial support from NSERC of Canada and H. Wu from the Univ. of Guelph. B.N. Di Stefano acknowledges full financial support from Nuptek Systems Ltd. The authors thank The Fields Institute for the hospitality while writing this paper.

References

1. Yuan, J., Mills, K.: Monitoring the Macroscopic Effect of DDoS Flooding Attacks. IEEE Transactions on Dependable and Secure Computing 2(4), 1–12 (2005)
2. `http://www.theregister.co.uk/2002/10/23/feds_investigating_largest_ever_internet/` (accessed January 5, 2009)
3. Nucci, A., Banneman, S.: Controlled Chaos. In: IEEE Spectrum, pp. 43–48 (December 2007)
4. Lawniczak, A.T., Wu, H., Di Stefano, B.N.: DDoS attack detection using entropy of packet traffic in CA like data communication network model. In: Adamatzky, A., et al. (eds.) Automata-2008 Theory and Applications of Cellular Automata, pp. 573–584. Luniver Press, UK (2008)
5. Lawniczak, A.T., Gerisch, A., Di Stefano, B.N.: Development and Performance of Cellular Automaton Model of OSI Network Layer of Packet Switching Networks. In: 16th IEEE CCECE 2003 – CCGEI 2003, vol. 2, pp. 1409–1412 (2003)
6. Lawniczak, A.T., Gerisch, A., Di Stefano, B.: OSI Network-layer Abstraction: Analysis of Simulation Dynamics and Performance Indicators. In: Mendes, J.F., et al. (eds.) AIP Conference Proc., New York, vol. 776, pp. 166–200 (2005)
7. Gerisch, A., Lawniczak, A.T., Di Stefano, B.: Building Blocks of a Simulation Environment of the OSI Network Layer of Packet Switching Networks. In: 16th IEEE CCECE 2003 – CCGEI 2003, p. 4 (2003)
8. Lawniczak, A.T., Gerisch, A., Maxie, K., Di Stefano, B.: Netzwerk: Migration of a Packet Switching Network Simulation Enviroment from MS Windows PC to Linux PC and to HPC. In: 19th International Symposium no High Performance Computing Systems and Applications, pp. 280–286. IEEE Press, Los Alamitos (2005)
9. Leon-Garcia, A., Widjaja, I.: Communication Networks: Fundamental Concepts and Key Architectures. McGraw-Hill Companies, Inc., New York (2000)

Identification of Social Tension in Organizational Networks

Relating Clan/Clique Formation to Social Tension

Ben Collingsworth and Ronaldo Menezes

Abstract. Analysis of email networks reveals properties similar to classic social networks such as homophily (assortativity) and community formation. The technology underlying email enables the formation of a network but it does not explain characteristics of the network that occur only as a result of patterns in human social behavior. Accordingly, a network formed from email activity correlates to the social environment and the dynamics of the environment used to create the network. Furthermore, the overall social behavior observed in an organization may be attributed directly to the organization's strength and resilience. When an organization is in trouble, we observe social tension among its employees. That being the case, one should be able to discern this tension by examining properties of a social network of the employees—the network should reflect the employees' mood; the fears, worries, gossips that are circulating, the good and the bad, are all reflected in organization's social network. One of the best representations of the true social organizational social network can be constructed from email exchange. The issue we investigate in this paper relates to timing: when does the network exhibits social tension when it is known to be present in the organization? In this paper, we provide a temporal analysis of the email social network constructed for the Enron Corporation and show that changes in network characteristics strongly correlate to real-world events in that organization. More importantly, we show that this correlation is time-shifted and appears in the network before the event becomes common knowledge; our hypothesis is that we can use the anomalies in the network to identify social tension in the organization and consequently help mitigate its consequences.

Ben Collingsworth
Bio-Inspired Computing Lab, Department of Computer Sciences, Florida Institute of Technology, Melbourne, Florida, USA
e-mail: bcolling@fit.edu

Ronaldo Menezes
Bio-Inspired Computing Lab, Department of Computer Sciences, Florida Institute of Technology, Melbourne, Florida, USA
e-mail: rmenezes@cs.fit.edu

R. Menezes et al. (Eds.): Complex Networks: CompleNet 2009, SCI 207, pp. 209–223.
springerlink.com © Springer-Verlag Berlin Heidelberg 2009

1 Introduction

The area of social sciences and organization engineering has improved our ability to identify problems within organizations. However, these approaches are limited as they generally rely on self-reports that are known have questionable reliability: *(i)* they depend on individuals telling the true, *(ii)* they depend on individuals having the correct view of the situation, and *(iii)* the approach itself does not scale well to organizations with a large number of employees. In today's market-driven world, governments and investors are constantly looking for tools to allow them to estimate the stability of organizations. Tools to perform organizational analysis and auditing are just a few of the many ways one can try to understand the health status of an organization [12] but these approaches are falling short in giving us the necessary confidence level in the results. It is not uncommon for an organization to be performing poorly without raising suspicion to investors, governments and other interested parties (including sectors of the organization itself). The best example can be the current global economic crisis that is affecting many industries that were thought to be sound.

From a social network perspective, we can overcome the aforementioned drawbacks because the network is a self-organized picture of the organization as does not depend on any particular individual collaborating or being truthful. In fact, the social network formed by individuals of network tend to be quite robust to localized or small-scale changes. But how can we get a social network of an organization? The answer to this question is not obvious and there may be many answers on how to do it. In the process of building social networks, one has to be careful to consider factors that could cause the network to be an ill-representation of the organization. For instance, one could get the subnetwork on sites such as Facebook or LinkedIn and consider that as the organization's network. Although possible, there are at least two problems with this approach. The first is that not all employees may be part of the network or be interested in being part of the network, and the second is that in these voluntary networks people are aware of social capital, meaning that they understand the value of having many connections, resulting in a network that rarely reflects the reality of the organization or the reality of the individuals' connections.

A better alternative to creating the organizational social network is email because it has become pervasive at all levels of organizations and *de facto* standard in communication; it is no longer voluntary in many organizations to have and use email. In the study described in this paper, we have used a database of the Enron email messages [6] and performed some analysis in the network formed by message exchange in an attempt to find anomalies in the network correlated with real, important events that took place during the less than two years of logs available in the database. Most people remember that in 2001 we had one of the biggest scandals in the history of public trading. Enron, an energy company primarily based in Houston, USA, went bankrupt after a series of irregular accounting procedures. This event took a lot of people by surprise and some insisted in denying to see the obvious: the company was going bankrupt. We believe that using approaches such as the one described in this paper we can have an early indication of tension within organizations.

It is worth noticing that, in the context of organizations, the network formed from email exchange is self-organized (not driven by any single force) and hence tends to be quite robust. Given this self-organization, anomalies become indeed interesting because the social networks have the tendency to self regulate against individual random acts. In the literature, this characteristic is referred to as the argument that social networks are resilient to random attacks due to their scale-free properties [1].

Organizational health may be viewed as the organization's ability to maintain employee engagement in productive task-oriented activities. A loss of health is indicated by a shift from a culture of employee commitment and productivity to a culture of self preservation driven by tension, uncertainty and fear. This shift may be the result of employee awareness of dysfunction within an organization. We argue that the change in employee behavior is detectable in email networks and we demonstrate this using primarily an analysis of clique and clan formation in a social network. Further, the network anomalies precede the outcome of the underlying dysfunction, and its detection may be used as an early-warning system in organizational assessment. Early crisis detection and intervention are critical in mitigating potentially catastrophic financial failures into recoverable downturns. It is also very important to understand that the correlation described in this paper *does not imply causality of the events*. This paper does nor argue that the tension caused the downturn of the Enron; we claim that the network already demonstrated the consequences of the downturn before it was clear to the general public.

Lastly, we argue that the work we describe here can be used not only as a prediction but also as an analysis tool to help us understand changes that could be implemented in the organization. Most organizations are used to having surveys and distributing questionnaires to employees asking about their satisfaction in the job. We argue that a more effective tool (or maybe a complementary one) would be to measure the effect of changes in the organization by looking for changes in the network. If we are correct in our assumption, network analysis can become a great asset on measuring the effectiveness of changes done within the organization.

We start this paper in Section 2 with a general description of complex networks and some of its properties that are used in this paper. We follow with Section 3 where we look at some of the major works done in this field and highlight how our approach differs from what exists in the literature. In Section 4, we provide the details on the construction of the Enron social network followed by our main results in Section 5. We then finish the paper describing possible research avenues that follow from the work described in this paper.

2 Complex Networks Concepts

The field of complex networks consists on the study of non-trivial topological characteristics of networks. In recent years, we have witnessed the growth of techniques to allow us to analyze networks in a holistic way rather than looking at the characteristics of subset of nodes (sub-graphs) or properties of individual nodes and edges. The ability to do large-scale statistical analysis of networks became possible with the advent of powerful computer systems capable of doing enormous amounts of computations in a very short amount of time [10].

Another field that has pushed the study of complex networks is the field of Social Sciences. A recent and pervasive phenomenon is the desire to understand interactions between people, or in other words study how people connect [2, 13]. Networks formed from human interactions are called *social networks*. The large social networks have non-trivial properties hence being characterized as complex. Some of the common properties of these networks are:

- **Degree Distribution:** Social networks have a heavy-tail degree distribution. This means that we have many nodes with few connections and a small number of nodes with a large number of connections.
- **Cliques:** The concept is quite simple and refers to a fully connected graph. That is, in a clique all individuals are connected to all other individuals in the graph via a direct connection. Cliques are an indication of the tightness of the network and are, in fact, strongly related to a measure in complex systems called clustering coefficient which is based on cliques of size three. This definition can be relaxed using the concept of a n-clique in which all elements of the clique are connected to each other by a path no longer than n (number of edges). Note that the basic definition of a clique refers to a 1-clique (henceforth just clique).
- **Clans:** Clans are also a measure of tightness. One can easily imagine that two individuals may be part of a 2-clique but the intermediate node that connects them not itself be part of the same clique. The clan concept takes care of this because a n-clan is defined as a structure in which all nodes must be connected to all other members of the clan using at most n intermediate edges but *all* intermediate nodes *must also be part of the same clan.*

Figure 1 shows a graph with an example of a clique and a 2-clan in that graph. In reality clans and cliques are not commonly used in complex network literature but rather in social networks. However since our complex network in this paper is also a social network the analysis based on cliques and clans is not out of the ordinary. In fact, cliques and clans have long been used in the study of social networks [9].

Other measures common in complex networks analysis are: network resilience, mixing patterns, hierarchical clustering, among others. However these do not apply directly to our study; an interested reader should refer to Newman [10] for a description of these and other properties of complex networks.

3 Related Work

A few papers in the literature investigate network properties of email networks. Their scale-free and small-world properties are shown in [5] and [11] for large university email networks. Both works discuss the issues associated with having incomplete data on email users external to the local email server. The clustering coefficient of the university email network is shown to be large compared to a random network of the same size. In addition, the mean shortest path is found to be consistent with

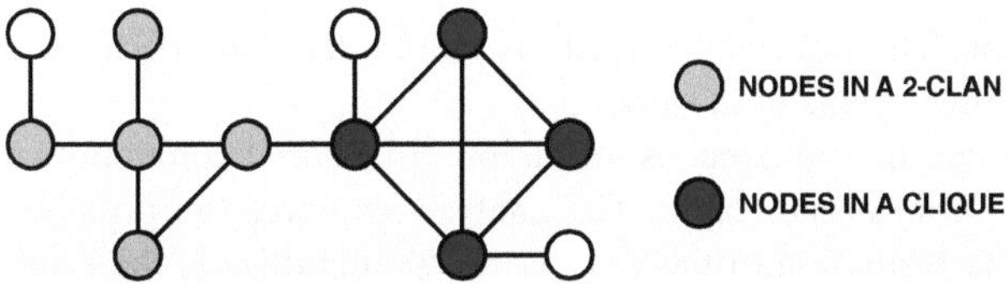

Fig. 1 The graph is marked with an example of a clique (dark colored nodes) and a 2-clan (light colored nodes). Nodes that are not colored are neither part of the clique or 2-clan displayed

small-world networks and much smaller than in random networks. Newman *et al.* [11] note the presence of reciprocity (the tendency of pairs of nodes to form mutual connections) in the email network several orders of magnitude larger than the reciprocity of a directed random network of similar size. The two works [5, 11] lead to discussions of the susceptibility of email networks to virus attack as seen in other scale-free networks. Immunization strategies are offered that take advantage of the scale free structure.

Similar to the analysis performed in this paper, Diesner *et al.* [4] examine communication patterns within Enron over time (as confirmed here in Figure 4). A variety of communication flows are analyzed including the number of messages exchanged within and between categories of employee position over time. Additionally, social network properties including network density, betweenness centrality, and strong component formation are studied. Several interesting anomalies are noted at the time of Enron's failure. Communications between board members diminished and nearly disappeared during the crisis. In contrast, managers and lawyers increased lateral communication and decreased upward communication throughout the crisis. A peak in average betweenness centrality is observed close to December 2001, suggesting network diversification and increased information flow at the time of failure. In addition, the number of strong components is shown to decrease throughout the crises, indicating a fracturing of the social network. The focus of Diesner is primarily on the analysis of Enron's failure (in and around December 2001). This paper extends the Diesner work by examining trends and anomalies leading up to the failure which may be used to identify and mitigate organizational issues before total failure occurs.

Zhou *et al.* [14] discern structure from the email communication history of employees. Similarly to this paper, they have worked with the Enron email network. Two algorithms are applied to determine organizational structure. The first algorithm looks at imbalance between senders and receivers. A node in a leadership position is presumed to receive more emails than it generates. A leadership score is assigned to each individual based on this imbalance. The second algorithm assigns a leadership score based on the radius of inclusion of a node in group emails. A node in a leadership position will have wider radius of inclusion. Organizational charts for Enron are produced by each of the algorithms. The organizational charts produced by Zhou *et al.* [14] appear to reflect informal structure rather than the formal hierarchy of an organization. The informal structure is more interesting because it is a true indication of how the organization works.

Guimerà *et al.* [7] examine the spontaneous formation of informal communities within the email network of an organization. The analysis explores the possibility that the force behind the formation of these informal communities is the same force at work in the formation of informal networks in other self-organized complex systems. A university email network containing 1700 users is analyzed. Properties such as cumulative community size distribution are calculated for the email network graph. The graph properties were found to be quite different from those generated from a random network. However, the properties are very similar to studies done on river-drainage system graphs. The underlying driving force in the evolution of river networks is energy optimization. The analysis concludes by suggesting that optimization may also be the force responsible for the formation of informal communities in an organization's email network. People form optimal communication networks through email in order to complete a cooperative task in an efficient manner.

4 Building the Enron Social Network

The characteristics of a social network can tell us quite a lot about the group that belong to the network. However this has not been done in the context of organizational stability. Our hypothesis is that the network can reveal information about the organization which may otherwise be undetectable by other individual-level analysis. To confirm our hypothesis we worked with the Enron email dataset [6], a well known archive of emails that were exchanged around the period of financial peril at Enron.

The Enron email logs were made available to the public allowing us to perform this study in the correlation between organizational tension and social network properties. The email logs span the period from energetic corporate growth (approx. August 2000) which sees a dramatic peak in stock value (approx. USD $90) to the complete corporate collapse of the corporation (approx. December 2001) when the stock price hit its lowest. During this period, the company culture shifted from an optimistic competitive atmosphere to a culture of fear, distrust, and cynicism. At the time of Enron's collapse, the corporate culture was pervaded with fear motivated by the threat of job loss, life savings, and reputation in the industry [3]. The Enron email dataset was downloaded from CMU University [6]. The data contained the messages associated with 147 unique email users. The downloaded messages came organized in folders according to user.

After the dataset was installed, each message file was parsed for fields of interest including the *From*, *To*, *Date*, *Time*, *Cc*, and *Bcc* message header fields. These fields were stored in a database table. One table entry was created for each message. Messages targeted to multiple recipients were expanded into multiple messages with a single recipient. The database was then pruned by discarding all messages that contained a *From* or *To* header field that did not have the `@enron.com` suffix. This reduction left only messages that were exchanged between Enron employees in the database. Additionally, the reduction eliminated junk email. The last refinement was to resolve email aliases within the Enron email dataset. The Levenshtein string distance algorithm [8] was used to group similar email addresses. A automated mechanism was applied to the sets of similar email addresses to reduce obvious aliases. The remaining groups of similar email addresses were resolved manually. Alias resolution resulted in creation of a database table containing a list of unique email participants. After the alias resolution was complete, occurrences of aliases in the *From* and *To* fields of the message database were replaced with the unique primary email address of the user.

Although the folders of only 147 Enron employees were released, around 14700 Enron employees are referenced in the Enron email database created. The large amount of additional users are the result of messages sent to and received from the core 147 employees to other employees at Enron. It should be noted that the full email history of these additional users have not been released. We are then left with two network configurations: a *core network* that contains only the 147 nodes and a *extended network* that contains all 14700 nodes. The extended network is not traditional because none of the nodes outside the 147 group have links between themselves (given we do not have their email logs). Figure 2 shows a view of the partial network.

Most of 147 users email folders released were senior management. The actual positions of about 75 percent of these people can be found a various Internet sites. This position data was incorporated in the data base as an attribute of each user.

All analysis of the Enron email was performed using the database created. Various database queries were performed to obtain desired data. For example, to obtain the in-degree of an employee for a particular month, a query was done requesting all messages containing a datastamp matching the desired month where the employee was identified as the recipient. The list returned from the database was then filtered so that each sender was counted only once. Network analysis tools (e.g. Pajek and

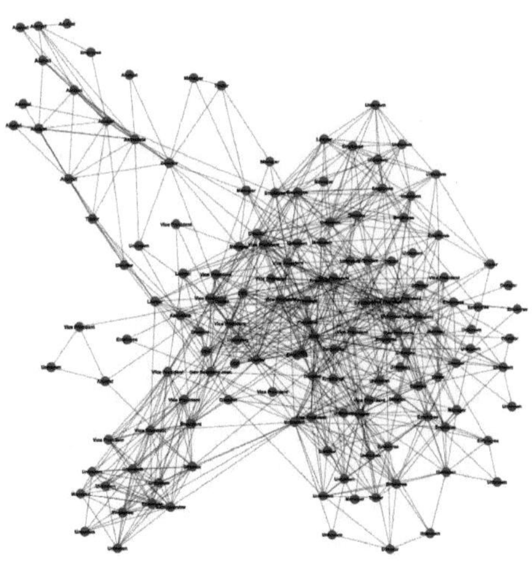

Fig. 2 Core Enron email network composed of 147 nodes

Cytoscape) were used to analyze the Enron data by extracting vertex and edge data, and attributes from the database and transforming this data into the format required by each tool.

5 Experimental Results

In spite of the incompleteness of the logs, we still can demonstrate properties of complex networks even though the expected characteristics may be outside the known ranges as described in Newman [10]. Recall that one of the properties of complex networks is that they are scale free, in other words $p_k \sim k^{-\alpha}$, meaning that the probability of node having a degree k follows a power law for some constant α. In Figure 3 we shows the cumulative degree distribution of the network close to bankruptcy (in December 2001). The figure also shows a trend line given by $p_k \sim k^{-\alpha_e}$, where $\alpha_e \approx 1.4 \pm 0.15$ for the extended network and $p_k \sim k^{-\alpha_c}$ for the core network, where $\alpha_c \approx 0.7 \pm 0.1$. According to earlier studies [10] the degree distribution of complex email networks the value the exponent, α,is in the range $[1.5, 2.0]$. Hence the Figure 3 is a strong indication that the networks we built for the Enron are structurally a social network.

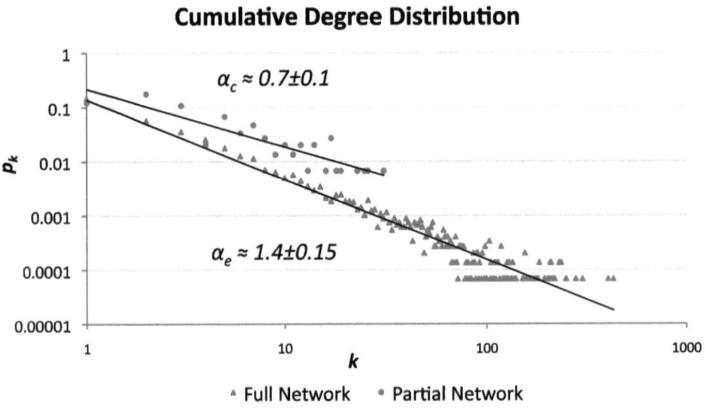

Fig. 3 log-log cumulative degree distribution of the extended and core networks close to time of bankruptcy

The core network refers to the 147 nodes for which we have all the folders as well as the representation of the other nodes which are inferred from the received and sent messages found in the 147 individuals represented in the logs. In this paper we will mostly look at a network that is formed by the 147 nodes; its cumulative degree distribution is also shown in Figure 3. Note that there is a weaker indication of a social network based on the α value (here 0.7 ± 0.1) but this is due to the network being small and hence more susceptible to variations.

The next step in out study was to confirm the main results found by Diesner *et al.* [4] in relation the average number of messages sent. Figure 4 shows that indeed the average number of messages sent decreases considerably around December 2001 hence confirming their studies and giving us more confidence on the network we build. Note however that small peak in communication around October 2001 meaning that during the crisis indeed communication ceased but not before that. Even more interesting is the peak around February 2001. What we can later in this section is that despite this increase in messages we do not see a correlation at that period in terms of clique and clan formation.

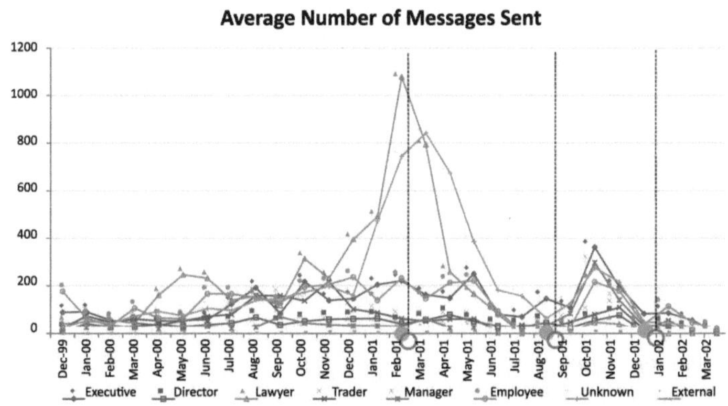

Fig. 4 Average number of messages sent per month by various employee positions

Now that we have seen that the Enron email logs form a social network, we can start looking more closely at other characteristics of this network. Since our goal is to find distress in the work environment due to stock prices falling we start with Figure 5 where we show the price of the stock against the number of messages circulating in the network and the number of participants.[1]

A close inspection of Figure 5 show us that there is a strong correlation between number of emails and participants. This is expected—more users tend to send more messages. Although at first glance there is no correlation between the stock price and the message volume and number of participants, with a bit more information about the history of Enron we can see that the network is demonstrating some stress at important points. Although several events happened during the period available in the logs it is clear from the literature on the history of the company [3] that three main events dominated:

February 2001: Jeffrey Skilling became the CEO on February 12. Kenneth Lay retains chairman post expecting a political appointment. Tom White resigns from EES (Enron Energy Services), the retail division he headed since 1998) and becomes secretary of the Army. He cashes out with 14 million dollars.

[1] Linear scaling in Figure 5 is used to aid visualization.

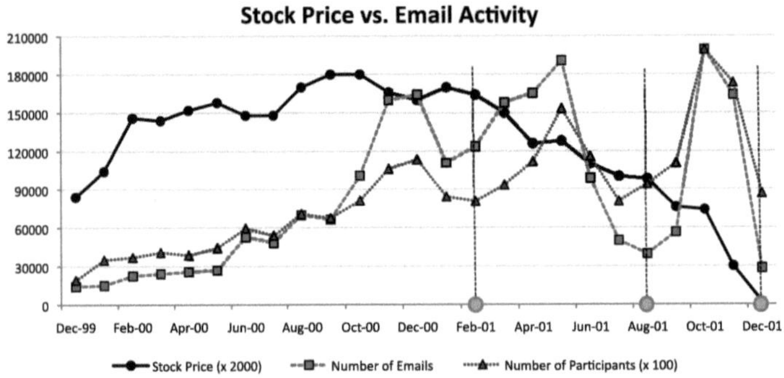

Fig. 5 Relation between message traffic (and number of participants) and the behavior of the stock price. The circles in the x-axis identify important events in the history of Enron

August 2001: On August 13 at a board meeting, Rick Buy outlines a disaster scenario if Enrons stock starts to fall—all SPEs (special purpose entities created to isolate financial risk) will crash. Jeffrey Skilling resigns next day after only six months as CEO, cites "personal reasons". Kenneth Lay re-takes CEO position on August 15. On August 21 Lay emails employees, stating *"one of my highest priorities is to restore investor confidence in Enron. This should result in a significantly higher stock price."*

December 2001: Enron files for chapter 11 bankruptcy December 2, Enron lays off 4000 employees.

If we now look at Figure 5 again and consider the dates above, we can see that the peaks happen just *before* the real events at an average of two months earlier; we argue that the network already demonstrated some emergent property indicating tension amongst the employees. This may be argued to be known and that when things were not going well at Enron it is clear that people will communicate more often. We are not claiming that this was unknown, what we claim here is that the network displays the phenomena and hence could be used in the future as a tool that can be analyzed in situations where the "problem" may not be yet known. E.g. finding about employee dissatisfaction.

Our main experiment in this paper is on the formation of cliques and clans and how this relates to the events identified above. It should be clear that the number of messages alone does not tell us to whom these messages were being sent and received. With the study of clans we want to demonstrate that not only the number of messages increased, but they did so in a very particular way: within clans/cliques. We have concentrated on two methods of clan formation. The first is to identify 2-clans; the other relates to assuming that employees in the same position also form a clan (henceforth called just position groups). Clearly we are using the name "clan" for the second case in a more social sciences way rather than the graph definition given earlier. In reality we want to observe homophily trends based on the position and clans—if we group individuals based on their position what is the behavior of in- and out-group, (positions) and in- and out-clans messages? The term in-clan message refers to message between dyads that are part of the same clan and out-clan refers to all the other messages (the same applies groups and cliques). Figure 6 shows the position clans while Figure 7 shows an example of the group found using the definition given earlier. The position groups shown in the figure are the ones used during the experiments that follow. What that means is that when we talk about in-group messages over time we are assuming the same groups during the entire execution of the experiment.

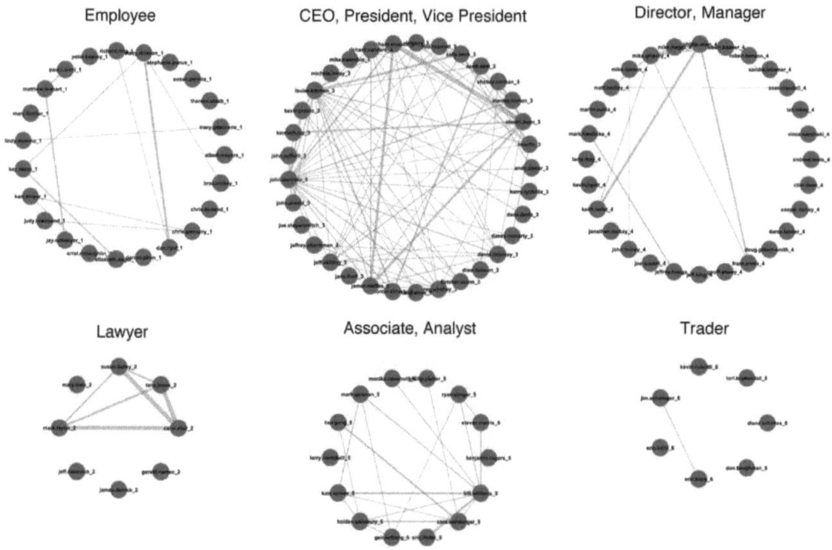

Fig. 6 Position clans formed based on a grouping of employees based on their job. This is a sample snapshot from April 2001

The question we want to answer then is what is the behavior of these clans, cliques and groups during the period of the email logs were available (spanning periods of high moral and periods of social tension). Does the network correlate the periods correctly? Recall that our hypothesis is that the tension caused by these events should be observed in the network before the event actually occurs. This should be the case because internally as the employees were probably already hearing rumors of such events even though the general public was unaware of them.

Figure 8 shows the evolution of the number of cliques in the network. Note that the clique formation does coincide with the identified events again with the same time-shift of 2 months. One may argue that clique formation may not be as indicative because it is restrictive. Hence we have also executed the same experiments but looking at clan formation, more specifically we looked at 2-clan formation. Figure 8 supports our hypothesis. A similar behavior to the clique formation can also be seen for clans. What is interesting is that the formation of cliques and clans demonstrate that the employees were increasingly talking to people forming closed groups.

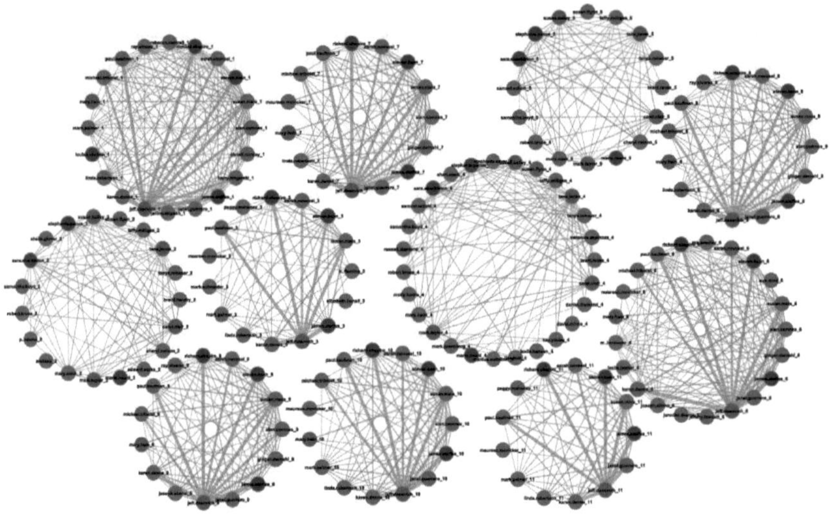

Fig. 7 2-clans formed based on the definition given earlier of employees based on their job. This is a sample snapshot from April 2001

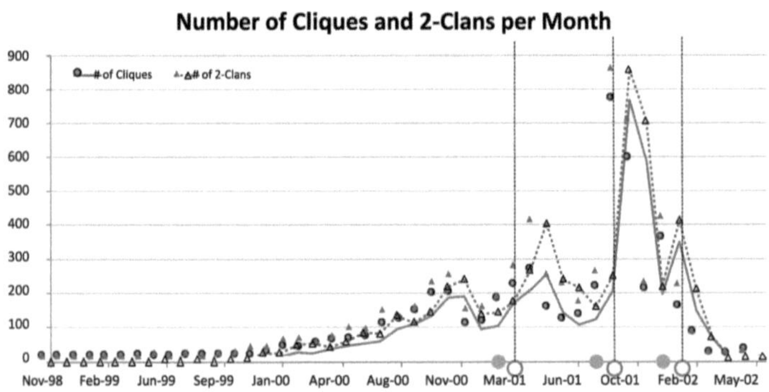

Fig. 8 We can confirm here that the peaks in the number of cliques in the graph take place also just before major events at Enron

Finally we look inside the clique and clan formation. What we are interested in seeing is whether the number of in-clan, in-clique and in-group formation also reflect the social tension. Note that cliques and clans can be formed but the messages between the dyads be outside the clan or clique. Figure 9 shows the behavior of in-clique, in-clan and in-group messages. Similarly, in-group messages are the ones between individuals in the same position in the organization.

Note that the main spikes in Figure 9 again occur a few months before the three events described earlier confirming our hypothesis that tension could be seen in the network. The number of messages sent/received shot up in a similar fashion as the

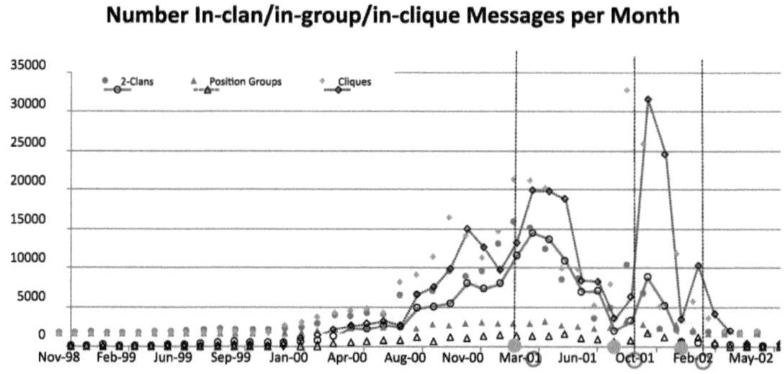

Fig. 9 The volume of messages not only increase during the period of tension but they increase within the clans/groups/cliques

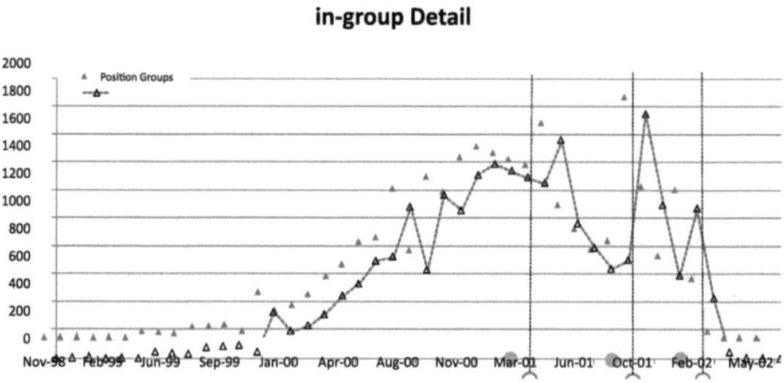

Fig. 10 The group increase does not correlate to the real events as strongly as with cliques and clans

general messages shown earlier. Given the increase on the number of messages in-clan and in-clique the in-group increase is as clear in Figure 9.

The increase of in-group messages, although not as prominent as in the in-clique and in-clan does occur in the position groups and can be seen in Figure 10. In fact, the only group that appear to have a significant increase is the group formed by high executives (CEO, President, vice-president). This result tell us that employees do not form tight groups based on their position in the company but most likely due to other affinity unrelated to their position.

6 Conclusion and Future Work

In this paper we have run several experiments on an email network and demonstrated that shifts in its emergent properties coincided with real-world events. We have used the email network from the Enron Corporation in the paper for several reasons: *(i)* its email logs are publicly available and span a relatively long period; *(ii)* there is a vast literature on the history of Enron that we can use to correlate the changes in the network with observations in the real world; and *(iii)* Enron is an interesting case because we saw the company going from being one of the most successful in the world to its complete collapse in a short amount of time while investors apparently knew little about what was going on internally.

What we argue here is that social network analysis can be used to complement tools to assess the stability of organizations. The network analysis can help in the identification of culture changes in the organization that may be hard to identify at the individual level. We go beyond the "culture change" and say that some shifts may be too prominent to be ignored. These networks tend to be quite robust, so changes in their emergent properties are a definite sign of employee tension/dissatisfaction or a sign of something major that may take place at the organization level.

As we discussed early in this paper, the motivation for this paper goes far beyond Enron which is used here as a case study. This paper reinforces early studies in that networks are able to reveal quite a lot about the organization. In our case, we can observe tension in the network with an increase in the number of tightly connected groups. We believe in the near future, organizations would be able to do real-time analysis of their social networks as a tool to measure many aspects in the organization such as employee satisfaction, acceptance of policy changes, etc. For instance, one can envision an organization implementing a policy and measuring the satisfaction of the policy among employees based on network analysis. In order to get to this stage other techniques would have to be developed allowing the real-time monitoring as well as mechanisms to isolate the changes related to the desired phenomenon.

We also would like to test our approach in other instances (other organizations). The problem is that very few of the cases that take place have email logs made available to the public. Currently in the USA we have seen signs of tension in very large companies due to a global financial crisis. Mortgage companies like Fannie Mae and Freddie Mac, and giants such as Ford, GM, and American Airlines are going through difficult times. Could investors have learned about these companies problems earlier through an analysis of their social networks? We strongly believe that to be the case. The organizations themselves can also benefit from tools based on social network analysis. Internal audits can be conducted to find such anomalies in an attempt to prevent more serious problems in the organization.

References

1. Albert, R., Jeong, H., Barabási, A.-L.: Error and attack tolerance of complex networks. Nature 406(6794), 378–382 (2000)
2. Barabási, A.-L.: Linked: The New Science of Networks. Perseus Publishing (2002)
3. Bryce, R.: Pipe Dreams: Greed, Ego, and the Death of Enron. PublicAffairs (October 2002)
4. Diesner, J., Frantz, T.L., Carley, K.M.: Communication networks from the enron email corpus: It's always about the people. enron is no different. Computational and Mathematical Organization Theory 11(3), 201–228 (2005)

5. Ebel, H., Mielsch, L.-I., Bornholdt, S.: Scale-free topology of e-mail networks. Phys. Rev. E 66(3), 035103 (2002)
6. Gervasio, M.: Enron email dataset (March 2004), `http://www.cs.cmu.edu/~enron/`
7. Guimerà, R., Danon, L., Díaz-Guilera, A., Giralt, F., Arenas, A.: Self-similar community structure in organisations. Phys. Rev. E 68, 065103 (2003)
8. Levenshtein, V.I.: Binary codes capable of correcting deletions, insertions, and reversals. Soviet Physics Doklady 10, 707–710 (1966)
9. Mokken, R.J.: Cliques, clubs and clans. Quality and Quantity 13, 161–173 (1979)
10. Newman, M.: The structure and function of complex networks. SIAM Review 45, 167–256 (2003)
11. Newman, M.E.J., Forrest, S., Balthrop, J.: Email networks and the spread of computer viruses. Phys. Rev. E 66(3), 035101 (2002)
12. Sterman, J.: System dynamics modeling: Tools for learning in a complex world. California Management Review 43(4), 8–25 (2001)
13. Watts, D.: Six Degrees: The Science of a Connected Age. Norton (2003)
14. Zhou, D., Song, Y., Zha, H., Zhang, Y.: Towards discovering organizational structure from email corpus. In: ICMLA 2005: Proceedings of the Fourth International Conference on Machine Learning and Applications, pp. 279–284. IEEE Computer Society, Washington (2005)

Author Index

MIX
Papier aus verantwortungsvollen Quellen
Paper from responsible sources
FSC® C105338
FSC
www.fsc.org

If you have any concerns about our products,
you can contact us on
ProductSafety@springernature.com

In case Publisher is established outside the EU,
the EU authorized representative is:
Springer Nature Customer Service Center GmbH
Europaplatz 3, 69115 Heidelberg, Germany

Printed by Libri Plureos GmbH
in Hamburg, Germany